VOLUME ONE HUNDRED AND NINETY FOUR

METHODS IN CELL BIOLOGY

Methods for Monitoring Mitochondrial Functions During Pathogen Infection

SERIES EDITOR

Lorenzo Galluzzi
Fox Chase Cancer Center,
Philadelphia, PA, United States

METHODS IN CELL BIOLOGY

Methods for Monitoring Mitochondrial Functions During Pathogen Infection

Edited by

SAVERIO MARCHI

Marche Polytechnic University, Ancona, Italy

LORENZO GALLUZZI

Fox Chase Cancer Center, Philadelphia, PA, United States

Academic Press is an imprint of Elsevier
125 London Wall, London EC2Y 5AS, United Kingdom
525 B Street, Suite 1650, San Diego, CA 92101, United States
50 Hampshire Street, 5th Floor, Cambridge, MA 02139, United States

First edition 2025

ISBN: 978-0-323-99219-0
ISSN: 0091-679X

For information on all Academic Press publications
visit our website at https://www.elsevier.com/books-and-journals

Publisher: Zoe Kruze
Editorial Project Manager: Devwart Chauhan
Production Project Manager: A. Maria Shalini
Cover Designer: Arumugam Kothandan

Typeset by STRAIVE, India

Contents

Contributors

Anushka Agarwal
Department of Biochemistry, School of Life Sciences, University of Hyderabad, India

Ajaz Ahmad
Department of Anesthesiology, University of Illinois College of Medicine, Chicago, IL, United States

Sharmistha Banerjee
Department of Biochemistry, School of Life Sciences, University of Hyderabad, India

Carmen Buchrieser
Institut Pasteur, Université Paris Cité, Biologie des Bactéries Intracellulaires and CNRS UMR 6047, Paris, France

Gloria D'Achille
Microbiology Unit, Department of Biomedical Sciences and Public Health, Marche Polytechnic University, Ancona, Italy

Nada Dhaouadi
Department of Clinical and Molecular Sciences, Marche Polytechnic University, Ancona, Italy

Mariangela Di Vincenzo
Department of Clinical and Molecular Sciences, Marche Polytechnic University, Ancona, Italy

Mariatou Dramé
Institut Pasteur, Université Paris Cité, Biologie des Bactéries Intracellulaires and CNRS UMR 6047, Paris, France

Pedro Escoll
Institut Pasteur, Université Paris Cité, Biologie des Bactéries Intracellulaires and CNRS UMR 6047, Paris, France

Matteo Fabbri
Section of Legal Medicine, Department of Translational Medicine, University of Ferrara, Ferrara, Italy

Marine Fidelle
Gustave Roussy; INSERM U1015, Equipe Labellisée—Ligue Nationale contre le Cancer, Villejuif, France

Shingo Hayashida
Department of Pediatrics and Child Health, Nihon University School of Medicine, Tokyo, Japan

Guochang Hu
Department of Anesthesiology; Department of Pharmacology & Regenerative Medicine, University of Illinois College of Medicine, Chicago, IL, United States

Valerio Iebba
Gustave Roussy Cancer Campus, Villejuif, France

Kazunori Kanemaru
Department of Physiology, Nihon University School of Medicine, Tokyo, Japan

Paulraj Kanmani
Department of Anesthesiology, University of Illinois College of Medicine, Chicago, IL, United States

Oliver Kepp
Metabolomics and Cell Biology Platforms, Gustave Roussy Cancer Center, Université Paris Saclay, Villejuif; Centre de Recherche des Cordeliers, Equipe labellisée par la Ligue contre le cancer, Université Paris Cité, Sorbonne Université, INSERM U1138, Institut Universitaire de France, Paris, France

Guido Kroemer
Metabolomics and Cell Biology Platforms, Gustave Roussy Cancer Center, Université Paris Saclay, Villejuif; Centre de Recherche des Cordeliers, Equipe labellisée par la Ligue contre le cancer, Université Paris Cité, Sorbonne Université, INSERM U1138, Institut Universitaire de France; Institut du Cancer Paris CARPEM, Department of Biology, APHP, Hôpital Européen Georges Pompidou, Paris, France

Marion Leduc
Metabolomics and Cell Biology Platforms, Gustave Roussy Cancer Center, Université Paris Saclay, Villejuif; Centre de Recherche des Cordeliers, Equipe labellisée par la Ligue contre le cancer, Université Paris Cité, Sorbonne Université, INSERM U1138, Institut Universitaire de France, Paris, France

Caterina Licini
Department of Clinical and Molecular Sciences, Marche Polytechnic University, Ancona, Italy

Fabio Marcheggiani
Department of Clinical and Molecular Sciences, Marche Polytechnic University, Ancona, Italy

Saverio Marchi
Department of Clinical and Molecular Sciences, Marche Polytechnic University; Advanced Technology Center for Aging Research, IRCCS INRCA, Ancona, Italy

Monica Mattioli-Belmonte
Department of Clinical and Molecular Sciences, Marche Polytechnic University; Advanced Technology Center for Aging Research, IRCCS INRCA, Ancona, Italy

Jayashankar Medikonda
Department of Biochemistry, School of Life Sciences, University of Hyderabad, India

Toshio Miki
Department of Physiology, Nihon University School of Medicine, Tokyo, Japan

Krishnaveni Mohareer
Department of Biochemistry, School of Life Sciences, University of Hyderabad, India

Gianluca Morroni
Microbiology Unit, Department of Biomedical Sciences and Public Health, Marche Polytechnic University, Ancona, Italy

Ilaria Nunzi
Department of Clinical and Molecular Sciences, Marche Polytechnic University, Ancona, Italy

Monia Orciani
Department of Clinical and Molecular Sciences, Marche Polytechnic University, Ancona, Italy

Daniel Schator
Institut Pasteur, Université Paris Cité, Biologie des Bactéries Intracellulaires and CNRS UMR 6047; Sorbonne Université, Collège Doctoral, Paris, France

Kanmani Suganya
Department of Anesthesiology, University of Illinois College of Medicine, Chicago, IL, United States

Isamu Taiko
Department of Physiology, Nihon University School of Medicine, Tokyo, Japan

Chika Takano
Division of Microbiology, Department of Pathology and Microbiology; Department of Pediatrics and Child Health, Nihon University School of Medicine, Tokyo, Japan

Ai-Ling Tian
Metabolomics and Cell Biology Platforms, Gustave Roussy Cancer Center, Université Paris Saclay, Villejuif; Centre de Recherche des Cordeliers, Equipe labellisée par la Ligue contre le cancer, Université Paris Cité, Sorbonne Université, INSERM U1138, Institut Universitaire de France, Paris, France

Sriram Yandrapally
Department of Biochemistry, School of Life Sciences, University of Hyderabad, India

Laurence Zitvogel
Gustave Roussy; INSERM U1015, Equipe Labellisée—Ligue Nationale contre le Cancer; Center of Clinical Investigations for In Situ Biotherapies of Cancer (BIOTHERIS) INSERM, CIC1428, Villejuif; Faculté de Médecine, Université Paris-Saclay, Kremlin-Bicêtre, France

CHAPTER ONE

Labeling of mitochondria for detection of intercellular mitochondrial transfer

Isamu Taiko[a], Chika Takano[b,c], Shingo Hayashida[c], Kazunori Kanemaru[a], and Toshio Miki[a,*]

[a]Department of Physiology, Nihon University School of Medicine, Tokyo, Japan
[b]Division of Microbiology, Department of Pathology and Microbiology, Nihon University School of Medicine, Tokyo, Japan
[c]Department of Pediatrics and Child Health, Nihon University School of Medicine, Tokyo, Japan
*Corresponding author: e-mail address: miki.toshio@nihon-u.ac.jp

Contents

Abstract

The phenomenon of intercellular transfer of mitochondria has been reported and has attracted significant interest in recent years. The phenomena involve a range of physiological and pathological conditions, such as tumor growth, immunoregulation, and tissue regeneration. There is speculation on the potential restoration of cellular energy status through the transfer of healthy mitochondria from donor cells to cells with impaired mitochondria. Multiple mechanisms and routes of mitochondria transfer have been suggested, including direct cell-to-cell connections, extracellular vesicles, and cell fusion. However, there is limited understanding regarding the precise mechanisms

Methods in Cell Biology, Volume 194
ISSN 0091-679X
https://doi.org/10.1016/bs.mcb.2024.05.001

behind mitochondrial transfer, particularly the initiation signals and the associated processes. In order to explore these fundamental mechanisms of mitochondrial transfer, it is imperative to employ techniques that enable direct labeling of mitochondria. Here, we present a detailed methodology utilizing fluorescent protein tagging to visualize mitochondria. The molecular biological techniques applied in this study entail the precise localization of mitochondria with reduced cytotoxicity. This approach facilitates the direct observation of transferred mitochondria through fluorescent and confocal microscopy. The described method can be readily implemented in other mammalian cell types with few modifications, enabling the continuous monitoring of mitochondrial trafficking processes over an extended period.

1. Introduction

Human diseases are often linked to mitochondrial dysfunction, which can be brought on by many conditions including oxidative stress, inflammation, and pathogenic bacterial infections (Gorman et al., 2016). The problem that remains of restoring mitochondrial functionality persists due to the hurdles associated with repairing each damaged mitochondrion. Recent studies have demonstrated the feasibility of transferring mitochondria from donor cells that are in a healthy state to recipient cells that possess impaired mitochondria. These findings suggest that the horizontal transfer of mitochondria may have therapeutic potential for the restoration of mitochondrial function (Mohammadalipour, Dumbali, & Wenzel, 2020). Koyanagi et al. initially documented the possibility of mitochondrial transfer between endothelial cells and cardiomyocytes (Koyanagi, Brandes, Haendeler, Zeiher, & Dimmeler, 2005). In the following year, a study by Spees et al. documented the ability of human bone marrow mesenchymal stromal cells to reinstate aerobic respiration in mtDNA-depleted ρ0 A549 lung adenocarcinoma cells. These cells, which exhibit a deficiency in ATP production via oxidative phosphorylation, were found to regain this metabolic function with the transfer of mesenchymal stromal cell-derived mitochondria (Spees, Olson, Whitney, & Prockop, 2006). Consequently, an increasing amount of research indicates that other cell types, particularly tissue stem cells such as mesenchymal stromal cells, may provide mitochondria. One of the recent studies conducted by Brestoff et al. successfully identified the occurrence of mitochondrial translocation from adipocytes to macrophages in a live animal (Brestoff et al., 2021). The transfer of mitochondria has been hypothesized to occur through various mechanisms, including direct intercellular connection, extracellular vesicles, or cell fusion (Acquistapace et al., 2011; Islam et al., 2012). Furthermore, it has been observed that mitochondria can be transferred via tunneling nanotubes (TNTs) or membrane nanotubes.

These tubule-like structures establish a direct connection between distinct cells, facilitating the transfer of cytosolic components, including mitochondria (Pasquier et al., 2013). The phenomenon of mitochondria transfer via TNTs has the potential to facilitate the specific and directed transfer of mitochondria, hence offering potential benefits in the context of targeted cell delivery for regenerative medicine. Nonetheless, there exists a dearth of comprehensive knowledge pertaining to the exact mechanisms underlying mitochondrial transfer.

In order to explore the underlying mechanisms of intercellular mitochondrial transfer, it is crucial to utilize methodologies that enable the direct labeling of mitochondria. The utilization of chemical fluorescent dyes, such as Tetramethyl Rhodamine Methyl Ester (TMRM) or MitoTracker dyes, is a commonly employed technique for the purpose of tagging mitochondria. However, it is common for these dyes to exhibit leakage when used for identifying mitochondria, resulting in inadvertent staining of other organelles. Furthermore, it has been observed that the accumulation of light-activated TMRM in mitochondria might lead to the generation of harmful reactive oxygen species. Additionally, the covalent labeling of mitochondrial matrix proteins with MitoTracker has been found to have a negative impact on mitochondrial respiration (Sargiacomo, Stonehouse, Moftakhar, & Sotgia, 2021; Zielonka et al., 2017). Hence, in order to accurately monitor the motion of mitochondria in a state of normalcy, it is advisable to utilize a mitochondrial labeling methodology that does not necessitate the use of chemical dyes.

As an alternative to chemical fluorescent labeling, mitochondrial targeting signal (MTS)-fused genetically encoded fluorescent proteins are frequently utilized in the mitochondrial research (Rizzuto, Brini, Pizzo, Murgia, & Pozzan, 1995). Misgeld et al. demonstrated that the utilization of stably expressed genetically encoded fluorescent proteins facilitates the extended monitoring of mitochondria, even in vivo (Misgeld, Kerschensteiner, Bareyre, Burgess, & Lichtman, 2007). Moreover, the utilization of cell type-specific promoters enables the selective expression of fluorescent proteins exclusively in the intended cells.

Here, we describe a detailed protocol to track mitochondrial transfer from placental stem cells, amniotic epithelial (AE) cells, to hydrogen peroxide (H_2O_2) exposed HEK293T cells by labeling mitochondria with fluorescent proteins (Miki, 2011; Takano et al., 2021). Since healthy AE cells are a type of fetal cells, the AE cells are considered to have less damaged mitochondria and are suitable as donor cells. The recipient cells were exposed to H_2O_2 because mitochondrial DNA is more susceptible than nuclear DNA to H_2O_2. Short-term exposure to an appropriate concentration of H_2O_2 can selectively damage mitochondrial DNA and thus induce mitochondrial dysfunction but not initiate cell death (Ballinger et al., 2000). This experimental setting particularly requires a method that relies on accurate targeting of mitochondria with less toxicity. Although some genetic labeling techniques mitochondrial targeted fluorescent proteins sometimes miss localized or leak to cytosol that causes cytotoxicity, our double tandem MTS repeat strategy successfully excludes any detectable miss localized signal and cytosolic leak. We have also selected fluorescent proteins with less toxicity in mitochondrial labeling by comparing several fluorescent proteins (Taiko et al., 2022). By labeling donor mitochondria and recipient cell cytosol with different colors, transferred mitochondria can be easily detected by confocal microscopy. To identify functional mitochondria, we also use a third color, MitoTracker Deep Red, which accumulates in mitochondria in a membrane potential-dependent manner. Our approach can be extended to directly monitor mitochondrial trafficking processes using long-term time-lapse imaging. Our method can be extended to direct monitoring of mitochondrial trafficking processes by long-term time-lapse imaging. With minimal variations, this method can be successfully implemented in a variety of human and other mammalian cell types.

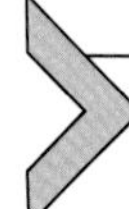

2. Materials

2.1 Common disposable

- Eppendorf Safe-Lock Tubes (#0030120086, Eppendorf) (see *Notes 1*)
- Falcon® 50 mL Conical Sterile Polypropylene Centrifuge Tubes (#352070, Corning) (see *Notes 1*)
- Falcon® 15 mL Conical Sterile Polypropylene Centrifuge Tubes (#352096, Corning) (see *Notes 1*)
- Nunc™ 100 × 15 mm Cell Culture Dish (#150466, Thermo Fisher Scientific) (see *Notes 1, 2*)
- Nunc™ Cell-Culture Treated Multidishes (24-well) (#142475, Thermo Fisher Scientific) (see *Notes 1*)
- Glass Bottom Dish (#D11530H, Matsunami Glass) (see *Notes 1*)

2.2 Cells and reagents

- HEK293T (#Q401, GenHunter) (see *Notes 1, 3*)
- Phosphate-buffered saline (PBS) (#166–23,555, FUJIFILM Wako Chemicals) (see *Notes 1, 3*)
- Dulbecco's Modified Eagle's Medium (DMEM), 4.5 g/L glucose (#045–30,285, FUJIFILM Wako Chemicals) (see *Notes 1, 3*)

- Fetal Bovine Serum (FBS) (#175012, NICHIREI BIOSCIENCES) (see *Notes 1, 3*)
- L-Glutamine (200 mM) (#25030081, Thermo Fisher Scientific) (see *Notes 1, 4*)
- Antibiotic-Antimycotic (100×) (#15240062, Thermo Fisher Scientific) (see *Notes 1, 5*)
- 0.5 g/L-Trypsin/0.53 mmol/L-EDTA (#32778-05, Nacalai Tesque) (see *Notes 1, 3*)
- 0.01% Collagen type I solution (#IFP9660, FUJIFILM Wako Chemicals) (see *Notes 1, 3*)
- Poly-L-lysine Hydrobromide (MW 70,000–150,000) (#28356-84, Nacalai Tesque) (see *Notes 1, 3*)
- Opti-MEM I Reduced Serum Medium (#31985070, Thermo Fisher Scientific) (see *Notes 1, 3*)
- PEI MAX—Transfection Grade Linear Polyethylenimine hydrochloride (MW 40,000) (#24765-100, Polysciences) (see *Notes 1, 6*)
- 1 mol/L-Sodium Hydroxide Solution (#37421-05, Nacalai Tesque) (see *Notes 1, 7*)
- pMD2. G plasmid (#12259, Addgene) (see *Notes 1, 8*)
- psPAX2 plasmid (#12260, Addgene) (see *Notes 1, 8*)
- pL-TurboRFPmt (Available on request) (see *Notes 8*) (Fig. 1)
- pLenti6-GFP (Available on request) (see *Notes 1, 8*)
- Lenti-X Concentrator (#631232 TAKARA BIO) (see *Notes 1, 3*)
- 30% *w/w* Hydrogen Peroxide (#081-04215, FUJIFILM Wako Chemicals) (see *Notes 1, 10*)
- MitoTracker Deep Red FM (#M22426, Thermo Fisher Scientific) (see *Notes 1, 3*)
- Dimethyl Sulfoxide (DMSO) (#D8418, Sigma-Aldrich) (see *Notes 1, 10*)

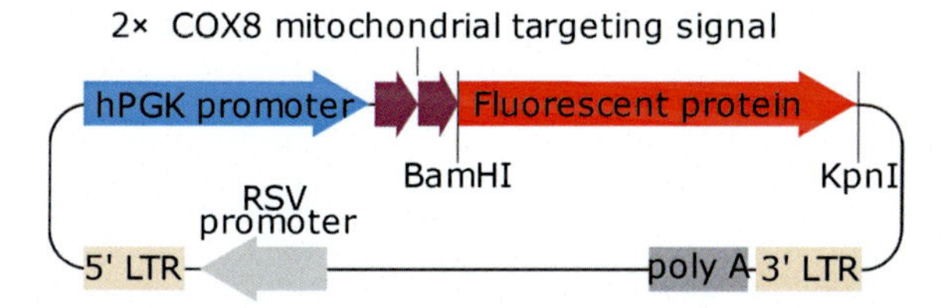

Fig. 1 The map of the lentiviral vector of the mitochondrial-targeted fluorescent protein described in Section 3.2. We employed human PGK promoter for mild expression of mitochondrial targeted fluorescent proteins. For strong mitochondrial targeting of fluorescent proteins, we tandemly inserted two mitochondrial targeting signals of human COX8 gene. Fluorescent proteins of interest can be inserted at BamHI and *KpnI* sites.

2.3 Equipment

- Automated cell counter, such as LUNA-II Automated Cell Counter (#L40002, Logos Biosystems) (see *Notes 1*)
- Humidified cell culture incubator, such as CytoGrow CO_2 Incubator (#MCO-50AIC, PHC Corporation) (see *Notes 1*)
- Laboratory biosafety cabinet (Class II), such as MHE-S901A2 (PHC Corporation) (see *Notes 1*)
- Standard centrifuge, such as Centrifuge 5810R (Eppendorf)
- Laboratory vortex mixer such as MixMate (#5353000529, Eppendorf) (see *Notes 1*)
- Laser scanning confocal microscope equipped with a 63× objective and lasers for 488, 555, 638 such as TCS SP8 (Leica Microsystems) (see *Notes 1*)

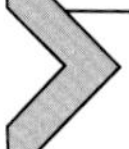

3. Methods

3.1 Reagent preparations

3.1.1 PEI (polyethylenimine) preparation

1. To prepare 2 mg/mL stock solution, 100 mg of PEI is dissolved in 45 mL deionized sterile water
2. pH of the solution is adjusted to 7.1 with 1 mol/L-Sodium Hydroxide Solution (see *Notes 7*)
3. Deionized sterile water is added to a final volume of 50 mL
4. PEI solution is filter-sterilized and store at 4 °C for up to 6 months.

3.1.2 Glass dish coating

1. To prepare 50 mg/mL stock solution, 5 mg of poly-L-lysine hydrobromide is dissolved in 100 mL deionized sterile water and store at 4 °C for up to 1 month. (see *Notes 11*)
2. 200 μL of 0.001% collagen and 50 mg/mL poly-L-lysine solution (1/10 for the collagen solution, 1/1000 for the poly-L-lysine stock solution) are added to the glass surface area of the glass bottom dishes followed by incubation for more than 30 min at room temperature. (see *Notes 12*)
3. The solution is removed and dishes are wash by deionized sterile water at two times and air dried for 5–10 min (see *Notes 13*)

3.2 Lentiviral production

3.2.1 Transfection for lentiviral production

1. The day before transfection, cells are seeded on collagen coated 10 cm dishes at a density of 5×10^5 with 10 mL of complete culture medium (see *Notes 14*)

2. For the transfection of one 10-cm dish, diluted DNA is prepared by mixing 12 mg of psPAX2, 6 mg of pMD2.G and 12 mg of pLenti6-GFP or pL-TurboRFPmt (or another lentiviral vector coding cytosolic or mitochondrial targeted fluorescent protein) to 500 μL of OMI (Opti-MEM I Reduced Serum Medium) in two 15 mL falcon tubes.
3. To prepare diluted PEI, 45 μL of 2 mg/mL PEI is diluted with 455 μL of OMI and mixed well (see *Notes 15*)
4. Transfection mix is prepared by dispensing 500 μL of diluted PEI to each tube of diluted DNA and mix well (see *Notes 15*)
5. After 15–20 min incubation, all 1000 μL of each transfection mix is added to each dish
6. After 6–12 h culture, medium is replaced by 10 mL of complete culture medium, followed by 48 h incubation at 37 °C under 5% CO_2

3.2.2 Collection of lentiviral particles

1. Each culture supernatant (10 mL) containing GFP or TurboRFPmt lentiviral particles is transferred to 50 mL Falcon tubes and centrifugated at 300 × *g* for 5 min at 4 °C
2. Clarified supernatants are recovered in new 50 mL Falcon tubes
3. Lenti-X Concentrator is added with 3 volumes of supernatant and mix well by gentle inversion followed by incubation for at least 30 min at 4 °C (see *Notes 16*)
4. Supernatants are centrifugated at 1500 × *g* for 45 min at 4 °C.
5. Supernatants are carefully removed, taking care not to disturb the pellet and the pellets are resuspended in 100 μL of PBS to obtain about 200 μL of viral suspension (see *Notes 17*)
6. This lentiviral suspension must be frozen at once before use or can be aliquoted and stored at −80 °C up to 1 month (see *Notes 18, 19*)

3.3 Lentiviral transduction

1. Both recipient HEK293T cells and donor AE cells or other cells are seeded on 24-well plate at a density of 50,000 cells/well the day before infection (see *Notes 20, 21*)
2. 20 μL of GFP or TurboRFPmt lentivirus suspension is added to the HEK293T cells or donor cells respectively, followed by 24–48 h incubation at 37 °C under 5% CO_2 (see *Notes 22*)
3. Exhausted culture medium is discarded by aspiration, and cells are gently washed with 500 mL PBS at 3 times and detached by trypsinization (see *Notes 23*)
4. Infected cells are propagated by seeding in new culture dishes

3.4 Mitochondria transfer

3.4.1 Recipient HEK293T cell seeding

1. Recipient GFP transduced HEK293T cells are detached by incubation in 2 mL 0.01% trypsin-EDTA solution at room temperature until cells are fully detached from the dish and collected in a 15 mL Falcon tube by adding 10 mL complete culture medium
2. Cells are pelleted by centrifuging the suspension at $300 \times g$ for 3 min, followed by aspiration of the supernatant and resuspension in 5 mL culture medium
3. Cells are counted on the automated Cell Counter and resuspended in culture medium to achieve a final concentration of 4×10^5 cells/mL
4. Cells are seeded on the glass area of collagen and poly-L-lysine coated dish at the density of 10,000 cells/, followed by 1–2 days incubation at 37 °C under 5% CO_2 (see *Notes 24*)

3.4.2 H_2O_2 treatment

1. 30% hydrogen peroxide is diluted by adding 300 mL of hydrogen peroxide to the 9.8 mL of culture medium to give a final concentration of 300 μM, immediately before use. (see *Notes 25–27*)
2. 2 days after HEK293T cell seeding, exhausted culture medium is discarded by aspiration
3. 2 mL of culture medium containing 300 μM hydrogen peroxide or control complete culture medium is gently added to the dishes
4. Cells are incubated at 37 °C for 1 h
5. Cells are gently and carefully washed with 2 mL PBS at three times
6. Donor TurboRFPmt transduced cells are detached by trypsinization, counted and resuspended in complete culture medium and stored on ice, immediately before end of 1 h H_2O_2 treatment of HEK293T cells.
7. Donor cells are seeded on the glass are of the dish at a density of 10,000 cells/dish

3.5 Image acquisition

3.5.1 Cell staining

1. To prepare 1 mM stock solution, 92 μL DMSO is added to MitoTracker tube and mix well
2. To prepare staining solution, 1 μL of 1 mM MitoTracker stock solution is diluted with 1 mL of complete culture medium in a 15 mL Falcon tube and mix well (see *Notes 28, 29*)

3. Exhausted culture medium from co-cultured recipient HEK293T and donor cells is discarded by aspiration, and cells are gently washed with 2 mL PBS
4. For MitoTracker loading, staining solution is added to the glass surface area of the dishes followed by incubation at 37 °C under 5% CO_2 for 10 min (see *Notes 30*)
5. After incubation, staining solution is carefully removed by pipetting and cells are gently washed with 2 mL PBS at least three times (see *Notes 31*)
6. 2 mL of complete culture medium is added to the cells

3.5.2 Image acquisition

1. Image acquisition is performed on laser scanning confocal microscopy TCS SP8 equipped with a 63× objective and excitation/emission wavelengths (nm) of 488/496–543 for GFP, 555/565–620 for TurboRFP, and 638/570–700 for MitoTracker Deep Red
2. Laser power setting is set to 0.5–1% and Images are captured at 400 Hz in 512 × 512 format (see *Notes 32*)
3. *Z*-stack images are acquired with capturing optical serial sections at 1 μm/slice
4. Images are analyzed by ImageJ software and 3D reconstructed images were obtained by Leica LAS X software (Fig. 2)

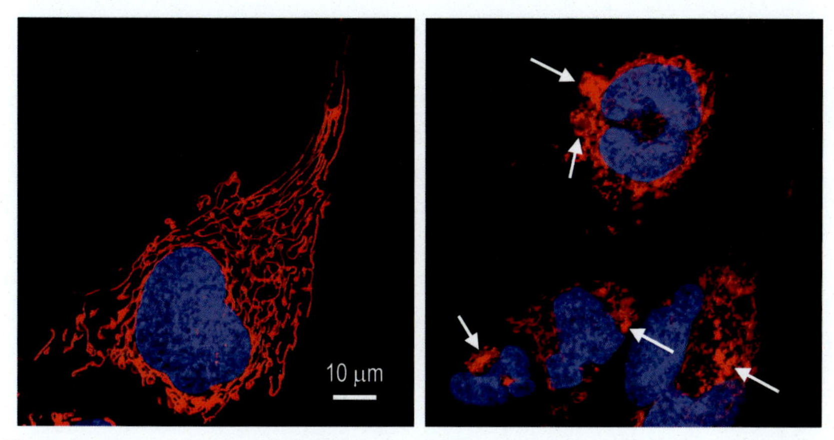

Fig. 2 Human amniotic epithelial derived cells were transuded with lentivirus harboring mitochondrial targeting TurboRFP as described in Section 3.3 (left) or higher viral MOI (right). Left: An ideal image of TurboRFP labeled mitochondria. Sharply defined mitochondrial outlines are shown. Right: A bad image of TurboRFP labeled mitochondria. Mitochondrial outlines are unclear and some of mitochondria are aggregated (white arrows) perhaps because excessive viral induction. Blue: Hoechst, red: TurboRFP.

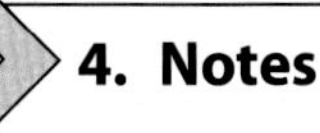

4. Notes

1. Catalog numbers and providers are provided as reference, but equivalent products can be purchased from a variety of sources at similar cost
2. Larger (15 cm) or smaller (6 cm) cell culture dishes can be employed should higher or lower amounts of cells be needed
3. HEK293T cells, as well as DMEM, FBS, opti-MEM, PBS, trypsin-EDTA, Collagen type I solution, Poly-L-lysine Hydrobromide and Lenti-X Concentrator are all considered as non-hazardous (NONH) but should be nonetheless manipulated with appropriate certified PPE
4. Glutamine are considered as hazards not otherwise classified (HNOC), and hence should be manipulated with appropriate certified PPE
5. Penicillin and streptomycin may cause an allergic skin reaction (H317), may cause allergy or asthma symptoms or breathing difficulties if inhaled (H334), and may damage fertility on the unborn child (H360), and hence should be manipulated by wearing appropriate certified PPE
6. PEI MAX can cause skin (H315) and eye (H320) irritation and hence it should be manipulated by wearing appropriate certified PPE
7. 1 mol/L-Sodium hydroxide solution can cause severe skin burns and eye damage (H314), serious eye damage (H318) and damage to respiratory system (H370), and hence should be manipulated by wearing appropriate certified PPE
8. Plasmids should be handled as a biohazardous material under Biosafety Level 2/Enhanced Biosafety Level 2 Containment while wearing appropriate certified PPE
9. Hydrogen peroxide can be toxic in contact with skin (H311) and fatal if inhaled (H330) and cause severe skin burns and eye damage (H314) and damage to respiratory system (H370), and hence should be manipulated with appropriate certified PPE
10. Dimethyl Sulfoxide may cause damage to respiratory system, and hence it should be manipulated by wearing appropriate certified PPE
11. For longer than 1 month, the stock solution must be stored at −20 °C
12. Prolonged coating time such as 24 h is no effect on cell culture.
13. Coated dishes can be stored at room temperature for at least 6 months
14. Cell density should be 50–80% confluent on the day of transfection
15. Reagents should be well mixed by vortex mixer for 30 s

16. Overnight incubation has less effect on virus recovery and titer
17. As necessary, culture medium can be used for pellets resuspension instead of PBS
18. Remaining viral producing cells may be responsible for false results
19. Repeated freeze-thaw cycles markedly reduce virus titer
20. Cells are susceptible to viral infection damage at low density
21. Culture scale can be smaller if necessary
22. Normally 24h incubation is sufficient for virus infection but full expression of the fluorescent proteins should require more than 48h culture
23. Remaining viral particle will stain recipient cell mitochondria leading to false results of mitochondrial transfer
24. Normally, 24h incubation is sufficient for downstream assay (H_2O_2 treatment)
25. Hydrogen peroxide has to be stored in light protected condition
26. The molarity of 30% hydrogen peroxide is about 9.8mol/L
27. Diluted hydrogen peroxide cannot be stored
28. To prevent aggregation, MitoTracker should be added slowly to the medium with vortex mixing
29. If PBS or other buffers are preferred 0.1% BSA should be supplemented
30. MitoTracker loading time must be less than 10min to prevent non-specific staining of other organellar
31. Remaining MitoTracker can cause higher background fluorescent signals and false staining of other organellar
32. Laser power should be set at least power as possible because too much laser power can induce mitochondrial disruption

5. Concluding remarks

The protocol described herein is a detailed workflow to detect mitochondrial transfer from healthy donor cells to H_2O_2-damaged cells by genetically labeling mitochondria with fluorescent proteins. Certain fluorescent proteins, such as DsRed and its derivatives, have been seen to occasionally exhibit notable lysosomal punctate aggregation. The utilization of our double tandem MTS repeat enables accurate subcellular targeting of fluorescent proteins to the mitochondria. Although this protocol employs confocal microscopy to visualize mitochondrial transfer, it can also be applicable to other fluorescence acquisition technologies, such as flow cytometry, without any major modification.

One primary constraint associated with this methodology is the unregulated alteration of genomic DNA in recipient cells due to the random integration of the lentiviral gene expression cassette. Due to the utilization of lentiviral vectors, there is a lack of control over the site of integration of the fluorescent protein gene into the genomic DNA. More precise genetic manipulation using CRISPR-Cas9, or other technologies will be required if this turns out to be a serious problem in the future. In this context, there remains potential for additional advancements in this methodology. Despite this and other limitations, the protocol detailed herein offers a simple and objective means to detect the mitochondrial transfer process.

Conflicts of interest

The authors declare no competing interests.

Appendix

Full sequences of lentiviral vector of mitochondria-targeted TurboRFP

TTGGGGTTGCGCCTTTTCCAAGGCAGCCCTGGGTTTGCG
CAGGGACGCGGCTGCTCTGGGCGTGGTTCCGGGAAACGCAG
CGGCGCCGACCCTGGGTCTCGCACATTCTTCACGTCCGTTC
GCAGCGTCACCCGGATCTTCGCCGCTACCCTTGTGGGCCCC
CCGGCGACGCTTCCTGCTCCGCCCCTAAGTCGGGAAGGTTC
CTTGCGGTTCGCGGCGTGCCGGACGTGACAAACGGAAGCCG
CACGTCTCACTAGTACCCTCGCAGACGGACAGCGCCAGGGA
GCAATGGCAGCGCGCCGACCGCGATGGGCTGTGGCCAATAG
CGGCTGCTCAGCAGGGCGCGCCGAGAGCAGCGGCCGGGAAG
GGGCGGTGCGGGAGGCGGGGTGTGGGGCGGTAGTGTGGGC-
CCTGTTCCTGCCCGCGCGGTGTTCCGCATTCTGCAAGCCTC
CGGAGCGCACGTCGGCAGTCGGCTCCCTCGTTGACCGAATC
ACCGACCTCTCTCCCCAGGGGGATCTACCGGTGTGACCATG
TCCGTCCTGACGCCGCTGCTGCTGCGGGGCTTGACAGGCTC
GGCCCGGCGGCTCCCAGTGCCGCGCGCCAAAATTCATTCAC
TGGGGGACCCCATGAGCGTGCTCACCCCACTCCTGCTGCGG
GGGCTGACCGGCAGCGCTAGGCGGCTGCCAGTCCCGCGGGC
CAAGATCCACAGTCTCGGCGATCCCGGATCCGCCACCATGAG
CGAGCTGATCAAGGAGAACATGCACATGAAGCTGTACATGGA
GGGCACCGTGAACAACCACCACTTCAAGTGCACATCCGAGGG

CGAAGGCAAGCCCTACGAGGGCACCCAGACCATGAAGATCAA
GGTGGTCGAGGGCGGCCCTCTCCCCTTCGCCTTCGACATCCT
GGCTACCAGCTTCATGTACGGCAGCAAAGCCTTCATCAACCA
CACCCAGGGCATCCCCGACTTCTTTAAGCAGTCCTTCCCTGA
GGGCTTCACATGGGAGAGAATCACCACATACGAAGACGGGGG
CGTGCTGACCGCTACCCAGGACACCAGCTTCCAGAACGGCTG
CATCATCTACAACGTCAAGATCAACGGGGTGAACTTCCCATC
CAACGGCCCTGTGATGCAGAAGAAAACACGCGGCTGGGAGGC
CAACACCGAGATGCTGTACCCCGCTGACGGCGGCCTGAGAGG
CCACAGCCAGATGGCCCTGAAGCTCGTGGGCGGGGGCTACCT
GCACTGCTCCTTCAAGACCACATACAGATCCAAGAAACCCGC
TAAGAACCTCAAGATGCCCGGCTTCCACTTCGTGGACCACAG
ACTGGAAAGAATCAAGGAGGCCGACAAAGAGACCTACGTCGA
GCAGCACGAGATGGCTGTGGCCAAGTACTGCGACCTCCCTAG
CAAACTGGGGCACAGATAAGGTACCTTTAAGACCAATGACTTA
CAAGGCAGCTGTAGATCTTAGCCACTTTTTAAAAGAAAAGGGG
GGACTGGAAGGGCTAATTCACTCCCAACGCTAGCAAGATCTG
CTTTTTGCTTGTACTGGGTCTCTCTGGTTAGACCAGATCTGAG
CCTGGGAGCTCTCTGGCTAACTAGGGAACCCACTGCTTAAGC
CTCAATAAAGCTTGCCTTGAGTGCTTCAAGTAGTGTGTGCCC
GTCTGTTGTGTGACTCTGGTAACTAGAGATCCCTCAGACCCT
TTTAGTCAGTGTGGAAAATCTCTAGCAGTAGTAGTTCATGTCA
TCTTATTATTCAGTATTTATAACTTGCAAAGAAATGAATATCA
GAGAGTGAGAGGAACTTGTTTATTGCAGCTTATAATGGTTACA
AATAAAGCAATAGCATCACAAATTTCACAAATAAAGCATTTTT
TTCACTGCATTCTAGTTGTGGTTTGTCCAAACTCATCAATGTA
TCTTATCATGTCTGGCTCTAGCTATCCCGCCCCTAACTCCGC
CCATCCCGCCCCTAACTCCGCCCAGTTCCGCCCATTCTCCGC
CCCATGGCTGACTAATTTTTTTTATTTATGCAGAGGCCGAGGC
CGCCTCGGCCTCTGAGCTATTCCAGAAGTAGTGAGGAGGCTT
TTTTGGAGGCCTAGTTCTAGAGTGGCCGGCTTTCCCCGTCAA
GCTCTAAATCGGGGGCTCCCTTTAGGGTTCCGATTTAGTGCT
TTACGGCACCTCGACCCCAAAAAACTTGATTAGGGTGATGGTT
CACGTAGTGGGCCATCGCCCTGATAGACGGTTTTTCGCCCTT
TGACGTTGGAGTCCACGTTCTTTAATAGTGGACTCTTGTTCCA
AACTGGAACAACACTCAACCCTATCTCGGTCTATTCTTTTGAT
TTATAAGGGATTTTGCCGATTTCGGCCTATTGGTTAAAAAATG
AGCTGATTTAACAAAAATTTAACGCGAATTTTAACAAAATATT
AACGCTTACAATTTAGGTGGCACTTTTCGGGGAAATGTGCGC

GGAACCCCTATTTGTTTATTTTTCTAAATACATTCAAATATGT
ATCCGCTCATGAGACAATAACCCTGATAAATGCTTCAATAATA
TTGAAAAAGGAAGAGTATGAGTATTCAACATTTCCGTGTCGCC
CTTATTCCCTTTTTTGCGGCATTTTGCCTTCCTGTTTTTGCTC
ACCCAGAAACGCTGGTGAAAGTAAAAGATGCTGAAGATCAGTT
GGGTGCACGAGTGGGTTACATCGAACTGGATCTCAACAGCGG
TAAGATCCTTGAGAGTTTTCGCCCCGAAGAACGTTTTCCAATG
ATGAGCACTTTTAAAGTTCTGCTATGTGGCGCGGTATTATCCC
GTATTGACGCCGGGCAAGAGCAACTCGGTCGCCGCATACACT
ATTCTCAGAATGACTTGGTTGAGTACTCACCAGTCACAGAAAA
GCATCTTACGGATGGCATGACAGTAAGAGAATTATGCAGTGCT
GCCATAACCATGAGTGATAACACTGCGGCCAACTTACTTCTGA
CAACGATCGGAGGACCGAAGGAGCTAACCGCTTTTTTGCACA
ACATGGGGGATCATGTAACTCGCCTTGATCGTTGGGAACCGG
AGCTGAATGAAGCCATACCAAACGACGAGCGTGACACCACGA
TGCCTGTAGCAATGGCAACAACGTTGCGCAAACTATTAACTG
GCGAACTACTTACTCTAGCTTCCCGGCAACAATTAATAGACT
GGATGGAGGCGGATAAAGTTGCAGGACCACTTCTGCGCTCG
GCCCTTCCGGCTGGCTGGTTTATTGCTGATAAATCTGGAGCC
GGTGAGCGTGGGTCTCGCGGTATCATTGCAGCACTGGGGCC
AGATGGTAAGCCCTCCCGTATCGTAGTTATCTACACGACGGG
GAGTCAGGCAACTATGGATGAACGAAATAGACAGATCGCTGA
GATAGGTGCCTCACTGATTAAGCATTGGTAACTGTCAGACCA
AGTTTACTCATATATACTTTAGATTGATTTAAAACTTCATTTT
TAATTTAAAAGGATCTAGGTGAAGATCCTTTTTGATAATCTCA
TGACCAAAATCCCTTAACGTGAGTTTTCGTTCCACTGAGCGT
CAGACCCCGTAGAAAAGATCAAAGGATCTTCTTGAGATCCTT
TTTTTCTGCGCGTAATCTGCTGCTTGCAAACAAAAAAACCAC
CGCTACCAGCGGTGGTTTGTTTGCCGGATCAAGAGCTACCAA
CTCTTTTTCCGAAGGTAACTGGCTTCAGCAGAGCGCAGATAC
CAAATACTGTTCTTCTAGTGTAGCCGTAGTTAGGCCACCACT
TCAAGAACTCTGTAGCACCGCCTACATACCTCGCTCTGCTAA
TCCTGTTACCAGTGGCTGCTGCCAGTGGCGATAAGTCGTGTC
TTACCGGGTTGGACTCAAGACGATAGTTACCGGATAAGGCGC
AGCGGTCGGGCTGAACGGGGGGTTCGTGCACACAGCCCAGC
TTGGAGCGAACGACCTACACCGAACTGAGATACCTACAGCGT
GAGCTATGAGAAAGCGCCACGCTTCCCGAAGGGAGAAAGGCG
GACAGGTATCCGGTAAGCGGCAGGGTCGGAACAGGAGAGCGC
ACGAGGGAGCTTCCAGGGGGAAACGCCTGGTATCTTTATAGT

CCTGTCGGGTTTCGCCACCTCTGACTTGAGCGTCGATTTTTG
TGATGCTCGTCAGGGGGGCGGAGCCTATGGAAAAACGCCAGC
AACGCGGCCTTTTTACGGTTCCTGGCCTTTTGCTGGCCTTTT
GCTCACATGTTCTTTCCTGCGTTATCCCCTGATTCTGTGGAT
AACCGTATTACCGCCTTTGAGTGAGCTGATACCGCTCGCCGC
AGCCGAACGACCGAGCGCAGCGAGTCAGTGAGCGAGGAAGC
GGAAGAGCGCCAATACGCAAACCGCCTCTCCCCGCGCGTTG
GCCGATTCATTAATGCAGCTGGCACGACAGGTTTCCCGACTG
GAAAGCGGGCAGTGAGCGCAACGCAATTAATGTGAGTTAGCT
CACTCATTAGGCACCCCAGGCTTTACACTTTATGCTTCCGGC
TCGTATGTTGTGTGGAATTGTGAGCGGATAACAATTTCACAC
AGGAAACAGCTATGACCATGATTACGCCAAGCGCGCAATTAA
CCCTCACTAAAGGGAACAAAAGCTGGAGCTGCAAGCTTAATG
TAGTCTTATGCAATACTCTTGTAGTCTTGCAACATGGTAACG
ATGAGTTAGCAACATGCCTTACAAGGAGAGAAAAAGCACCGT
GCATGCCGATTGGTGGAAGTAAGGTGGTACGATCGTGCCTTA
TTAGGAAGGCAACAGACGGGTCTGACATGGATTGGACGAACC
ACTGAATTGCCGCATTGCAGAGATATTGTATTTAAGTGCCTA
GCTCGATACATAAACGGGTCTCTCTGGTTAGACCAGATCTGA
GCCTGGGAGCTCTCTGGCTAACTAGGGAACCCACTGCTTAAG
CCTCAATAAAGCTTGCCTTGAGTGCTTCAAGTAGTGTGTGCC
CGTCTGTTGTGTGACTCTGGTAACTAGAGATCCCTCAGACCC
TTTTAGTCAGTGTGGAAAATCTCTAGCAGTGGCGCCCGAACA
GGGACTTGAAAGCGAAAGGGAAACCAGAGGAGCTCTCTCGAC
GCAGGACTCGGCTTGCTGAAGCGCGCACGGCAAGAGGCGAG
GGGCGGCGACTGGTGAGTACGCCAAAAATTTTGACTAGCGGA
GGCTAGAAGGAGAGAGATGGGTGCGAGAGCGTCAGTATTAAG
CGGGGGAGAATTAGATCGCGATGGGAAAAAATTCGGTTAAGG
CCAGGGGGAAAGAAAAAATATAAATTAAAACATATAGTATGGG
CAAGCAGGGAGCTAGAACGATTCGCAGTTAATCCTGGCCTGT
TAGAAACATCAGAAGGCTGTAGACAAATACTGGGACAGCTACA
ACCATCCCTTCAGACAGGATCAGAAGAACTTAGATCATTATAT
AATACAGTAGCAACCCTCTATTGTGTGCATCAAAGGATAGAGA
TAAAAGACACCAAGGAAGCTTTAGACAAGATAGAGGAAGAGCA
AAACAAAAGTAAGACCACCGCACAGCAAGCGGCCGCTGATCT
TCAGACCTGGAGGAGGAGATATGAGGGACAATTGGAGAAGTG
AATTATATAAATATAAAGTAGTAAAAATTGAACCATTAGGAGT
AGCACCCACCAAGGCAAAGAGAAGAGTGGTGCAGAGAGAAAA
AAGAGCAGTGGGAATAGGAGCTTTGTTCCTTGGGTTCTTGGG

AGCAGCAGGAAGCACTATGGGCGCAGCGTCAATGACGCTGAC
GGTACAGGCCAGACAATTATTGTCTGGTATAGTGCAGCAGCA
GAACAATTTGCTGAGGGCTATTGAGGCGCAACAGCATCTGTT
GCAACTCACAGTCTGGGGCATCAAGCAGCTCCAGGCAAGAAT
CCTGGCTGTGGAAAGATACCTAAAGGATCAACAGCTCCTGGG
GATTTGGGGTTGCTCTGGAAAACTCATTTGCACCACTGCTGT
GCCTTGGAATGCTAGTTGGAGTAATAAATCTCTGGAACAGATT
TGGAATCACACGACCTGGATGGAGTGGGACAGAGAAATTAAC
AATTACACAAGCTTAATACACTCCTTAATTGAAGAATCGCAAA
ACCAGCAAGAAAAGAATGAACAAGAATTATTGGAATTAGATAA
ATGGGCAAGTTTGTGGAATTGGTTTAACATAACAAATTGGCTG
TGGTATATAAAATTATTCATAATGATAGTAGGAGGCTTGGTAG
GTTTAAGAATAGTTTTTGCTGTACTTTCTATAGTGAATAGAGT
TAGGCAGGGATATTCACCATTATCGTTTCAGACCCACCTCCC
AACCCCGAGGGGACCCGACAGGCCCGAAGGAATAGAAGAAGA
AGGTGGAGAGAGAGACAGAGACAGATCCATTCGATTAGTGAA
CGGATCTCGACGGTATCGATCACGAGACTAGCCTCGACCTAG
GCAAATGGCAGTATTCATCCACAATTTTAAAAGAAAAGGGGGG
ATTGGGGGGTACAGTGCAGGGGAAAGAATAGTAGACATAATAG
CAACAGACATACAAACTAAAGAATTACAAAAACAAATTACAAA
AATTCAAAATTTTCGGGTTTATTACAGGGACAGCAGAGATCCA
CTTTGGCCGCGGATCGAGGGGG.

References

Acquistapace, A., Bru, T., Lesault, P.-F., Figeac, F., Coudert, A. E., Le Coz, O., et al. (2011). Human mesenchymal stem cells reprogram adult cardiomyocytes toward a progenitor-like state through partial cell fusion and mitochondria transfer. *Stem Cells (Dayton, Ohio)*, *29*(5), 812–824. https://doi.org/10.1002/stem.632.

Ballinger, S. W., Patterson, C., Yan, C.-N., Doan, R., Burow, D. L., Young, C. G., et al. (2000). Hydrogen peroxide and peroxynitrite-induced mitochondrial DNA damage and dysfunction in vascular endothelial and smooth muscle cells. *Circulation Research*, *86*(9), 960–966. https://doi.org/10.1161/01.RES.86.9.960.

Brestoff, J. R., Wilen, C. B., Moley, J. R., Li, Y., Zou, W., Malvin, N. P., et al. (2021). Intercellular mitochondria transfer to macrophages regulates white adipose tissue homeostasis and is impaired in obesity. *Cell Metabolism*, *33*(2), 270–282.e8. https://doi.org/10.1016/j.cmet.2020.11.008.

Gorman, G. S., Chinnery, P. F., DiMauro, S., Hirano, M., Koga, Y., McFarland, R., et al. (2016). Mitochondrial diseases. *Nature Reviews Disease Primers*, *2*(1), 1–22. https://doi.org/10.1038/nrdp.2016.80.

Islam, M. N., Das, S. R., Emin, M. T., Wei, M., Sun, L., Westphalen, K., et al. (2012). Mitochondrial transfer from bone marrow-derived stromal cells to pulmonary alveoli protects against acute lung injury. *Nature Medicine*, *18*(5), 759–765. https://doi.org/10.1038/nm.2736.

Koyanagi, M., Brandes, R. P., Haendeler, J., Zeiher, A. M., & Dimmeler, S. (2005). Cell-to-cell connection of endothelial progenitor cells with cardiac myocytes by nanotubes. *Circulation Research, 96*(10), 1039–1041. https://doi.org/10.1161/01.RES.0000168650.23479.0c.

Miki, T. (2011). Amnion-derived stem cells: In quest of clinical applications. *Stem Cell Research & Therapy, 2*(3), 25. https://doi.org/10.1186/scrt66.

Misgeld, T., Kerschensteiner, M., Bareyre, F. M., Burgess, R. W., & Lichtman, J. W. (2007). Imaging axonal transport of mitochondria in vivo. *Nature Methods, 4*(7), 559–561. https://doi.org/10.1038/nmeth1055.

Mohammadalipour, A., Dumbali, S. P., & Wenzel, P. L. (2020). Mitochondrial transfer and regulators of mesenchymal stromal cell function and therapeutic efficacy. *Frontiers in Cell and Developmental Biology, 8*, 1519. https://doi.org/10.3389/fcell.2020.603292.

Pasquier, J., Guerrouahen, B. S., Al Thawadi, H., Ghiabi, P., Maleki, M., Abu-Kaoud, N., et al. (2013). Preferential transfer of mitochondria from endothelial to cancer cells through tunneling nanotubes modulates chemoresistance. *Journal of Translational Medicine, 11*(1), 94. https://doi.org/10.1186/1479-5876-11-94.

Rizzuto, R., Brini, M., Pizzo, P., Murgia, M., & Pozzan, T. (1995). Chimeric green fluorescent protein as a tool for visualizing subcellular organelles in living cells. *Current Biology, 5*(6), 635–642. https://doi.org/10.1016/S0960-9822(95)00128-X.

Sargiacomo, C., Stonehouse, S., Moftakhar, Z., & Sotgia, F. (2021). MitoTracker Deep Red (MTDR) is a metabolic inhibitor for targeting mitochondria and eradicating cancer stem cells (CSCs), with anti-tumor and anti-metastatic activity in vivo. *Frontiers in Oncology, 11*. https://www.frontiersin.org/article/10.3389/fonc.2021.678343.

Spees, J. L., Olson, S. D., Whitney, M. J., & Prockop, D. J. (2006). Mitochondrial transfer between cells can rescue aerobic respiration. *Proceedings of the National Academy of Sciences, 103*(5), 1283–1288. https://doi.org/10.1073/pnas.0510511103.

Taiko, I., Takano, C., Nomoto, M., Hayashida, S., Kanemaru, K., & Miki, T. (2022). Selection of red fluorescent protein for genetic labeling of mitochondria and intercellular transfer of viable mitochondria. *Scientific Reports, 12*(1), 19841. https://doi.org/10.1038/s41598-022-24297-0.

Takano, C., Grubbs, B. H., Ishige, M., Ogawa, E., Morioka, I., Hayakawa, S., et al. (2021). Clinical perspective on the use of human amniotic epithelial cells to treat congenital metabolic diseases with a focus on maple syrup urine disease. *Stem Cells Translational Medicine, 10*(6), 829–835. https://doi.org/10.1002/sctm.20-0225.

Zielonka, J., Sikora, A., Hardy, M., Ouari, O., Vasquez-Vivar, J., Cheng, G., et al. (2017). Mitochondria-targeted triphenylphosphonium-based compounds: Syntheses, mechanisms of action, and therapeutic and diagnostic applications. *Chemical Reviews, 117*(15), 10043–10120. https://doi.org/10.1021/acs.chemrev.7b00042.

CHAPTER TWO

Visualizing mitochondrial electron transport chain complexes and super-complexes during infection of human macrophages with *Legionella pneumophila*

Mariatou Dramé[a], Daniel Schator[a,b], Carmen Buchrieser[a,*], and Pedro Escoll[a,*]

[a]Institut Pasteur, Université Paris Cité, Biologie des Bactéries Intracellulaires and CNRS UMR 6047, Paris, France
[b]Sorbonne Université, Collège Doctoral, Paris, France
[*]Corresponding authors: e-mail address: cbuch@pasteur.fr; pescoll@pasteur.fr

Contents

Methods in Cell Biology, Volume 194
ISSN 0091-679X
https://doi.org/10.1016/bs.mcb.2024.01.003

Abstract

The ultrastructure of mitochondria is pivotal for their respiratory activity. Thus, the regulation of the assembly of the super-complexes (SCs) of the mitochondrial electron transport chain (ETC) might be a core aspect of macrophage immunometabolism during bacterial infection. In order to study the impact of infection by *Legionella pneumophila* on the configuration of mitochondrial complexes and SCs in human macrophages, we have adapted and combined different methods such as cell sorting of infected cells, magnetic isolation of highly pure and functional mitochondria, quality control of mitochondrial purity by flow cytometry, and BN-PAGE (Blue-Native Polyacrylamide Gel Electrophoresis) coupled to Western Blot using near-infrared (NIR) fluorescence. The here presented protocol uses infected and non-infected human macrophage-like THP-1 cells and GFP-expressing *L. pneumophila*, but the method can be used to analyze the configuration of ETC complexes and SCs also in other mammalian cells and infected with different intracellular bacteria expressing a fluorescent protein.

1. Introduction

Macrophages are professional phagocytic cells at the first line of the immune defense against pathogenic microorganisms. They are present in all tissues and hence show a large heterogeneity and functional diversity related to the regulation of homeostasis, such as host defense, wound healing and general immune regulation (Mosser & Edwards, 2008). Therefore, macrophages are extremely plastic and flexible cells, and their different functions largely depend on their metabolic status.

Macrophage metabolism is at the center of their immune responses. Key players of the cellular metabolism and metabolic reprogramming of macrophages are mitochondria (Van den Bossche, O'Neill, & Menon, 2017). Besides their role in cellular bioenergetics, mitochondria are eukaryotic organelles involved in many other cellular processes such as calcium homeostasis, amino acids synthesis, or programmed cell death (West, Shadel, & Ghosh, 2011). Importantly, the mitochondrial metabolism is also highly implicated in innate immunity signaling pathways against viral and bacterial pathogens. The tricarboxylic acid (TCA) cycle and the oxidative phosphorylation (OXPHOS), both taking place inside mitochondria, play essential roles in the metabolic alterations triggered in macrophages after they encounter pathogen-associated molecular patterns (PAMPs) such as bacterial lipopolysaccharides (LPS) (Martínez-Reyes & Chandel, 2020; O'Neill & Pearce, 2016). After activation, alterations in the TCA cycle and OXPHOS in macrophages lead, through several different pathways, to the expression and secretion of the pro-inflammatory cytokine Interleukin-1β (IL-1β), to the production of mitochondrial reactive oxygen species (mROS) and of itaconate, which holds antimicrobial properties (Dramé, Buchrieser, & Escoll, 2020; O'Neill & Pearce, 2016). In addition, perturbations of glycolysis or OXPHOS have also been shown to activate the NOD-like receptor pyrin 3 (NLRP3) inflammasome formation. NLRP3 inflammasomes are in vicinity of mitochondria to sense mitochondrial dysfunctions, and their activation in macrophages leads to the production of IL-1β and eventually to pyroptosis, a pro-inflammatory form of programmed cell death triggered in macrophages (Dramé et al., 2020; O'Neill & Pearce, 2016; Zhou, Yazdi, Menu, & Tschopp, 2011). Thus, current knowledge supports a model where mitochondria have a prominent role as signaling hubs during bacterial infection, as they coordinate several anti-bacterial responses of macrophages, such as cytokine production, the generation of anti-microbial molecules such as itaconate, as well as cell death/pyroptosis.

A pivotal structure in mitochondria, and the platform where OXPHOS takes place, is the electron transport chain (ETC). It is constituted of five complexes referred to as Complex I to Complex V which are positioned at the cristae of the mitochondrial inner membrane. Released electrons coming from the oxidation of organic compounds during glycolysis and the TCA cycle are transported in the form of reduced Nicotinamide Adenine Dinucleotide (NADH) and Flavin Adenine Dinucleotide ($FADH_2$) into the ETC. Oxidation of NADH and $FADH_2$ to NAD+ and FAD+ by Complex I and Complex II, respectively, release these electrons into the ETC. The flow of electrons through Complex I, Complex II and Complex IV provokes protons (H^+) to be concomitantly translocated through these complexes to the intermembrane space (IMS) of mitochondria, which creates a mitochondrial membrane potential ($\Delta\Psi_m$). At Complex IV, electrons are accepted by the final electron acceptor O_2, forming H_2O. Finally, as protons accumulate in the IMS, a specific form of potential energy is created, the proton motive force. Then Complex V (the F_OF_1-ATPase) uses the kinetic energy provided by the proton motive force to create ATP as protons pass through it (Nolfi-Donegan, Braganza, & Shiva, 2020). The organization of these complexes has been debated for a long time, contrasting a "fluid model" where individual complexes freely move in the mitochondrial inner membrane, and a "solid model" where complexes are all joined in one big structure (Lenaz & Genova, 2007). However, in 2008, Acín-Pérez and colleagues proposed a new model called the "plasticity model," which is generally accepted today. In this model, complexes can indeed associate between themselves with varying stoichiometries, forming super-complexes (SCs) but also exist as individual complexes (Acín-Pérez, Fernández-Silva, Peleato, Pérez-Martos, & Enriquez, 2008).

Findings by Garaude et al. showed that infection of murine macrophages with extracellular, living bacteria (*Escherichia coli*) transiently decreases the assembly of the ETC Complex I and Complex I-containing SCs and switched the relative contributions of Complex I and Complex II to overall mitochondrial respiration (Garaude et al., 2016). These adaptations of ETC SCs in macrophages were driven by a signaling pathway that includes phagosomal NADPH oxidase, the reactive oxygen species (ROS)-dependent tyrosine kinase Fgr, Toll-like receptors and the NLRP3 inflammasome (Garaude et al., 2016). Pathogenic bacteria such as *Legionella pneumophila, Salmonella enterica* or *Mycobacterium tuberculosis* infect human macrophages where they can grow intracellularly (Mitchell, Chen, & Portnoy, 2016). As they replicate within host cells, intracellular bacteria can obtain resources for bacterial growth only from the host cell they infect. Therefore, it has been suggested that bacteria-induced alterations of the macrophage metabolism during infection, in particular OXPHOS, might benefit pathogenic bacteria by redirecting cellular resources, such as glycolytic flux or TCA cycle intermediates, to biosynthetic pathways that might sustain intracellular bacterial replication (Escoll & Buchrieser, 2018; Marchi, Morroni, Pinton, & Galluzzi, 2022; Tiku, Tan, & Dikic, 2020). Therefore, the adaptation of the ETCs might differ during infection with extracellular and intracellular bacteria. We have previously shown that *L. pneumophila*, the causative agent of Legionnaires' disease in humans, alters mitochondrial morphology during infection of macrophages by translocating the bacterial effector MitF in the host cell through its type IV secretion system (T4SS), thereby reducing OXPHOS and ATP production in infected human macrophages (Escoll et al., 2017). Furthermore, we recently found that *L. pneumophila* also modulates the activity of ETC Complex V in a T4SS-dependent manner which allows to preserve $\Delta\Psi_m$ of infected human macrophages and to delay macrophage cell death (Escoll et al., 2021). Thus, *L. pneumophila* has evolved molecular tools to manipulate mitochondrial bioenergetics (García-Rodríguez, Buchrieser, & Escoll, 2023).

The ultrastructure of mitochondria is pivotal for their respiratory activity (Cogliati, Enriquez, & Scorrano, 2016) and murine macrophages regulate SCs during infection with extracellular bacteria (Garaude et al., 2016), thus we hypothesized that the regulation of SC assembly might be a core aspect of macrophage immunometabolism during bacterial infection. To analyze whether the infection by intracellular bacteria impacts the configuration of mitochondrial SCs in human macrophages, we combined cell sorting of infected cells, magnetic isolation of highly pure and functional mitochondria, quality control of mitochondrial purity by flow cytometry, and BN-PAGE coupled to Western Blot using near-infrared (NIR) fluorescence to analyze mitochondrial ETC complexes and SCs of infected cells (Fig. 1). The protocol was setup using infected and non-infected human macrophage-like

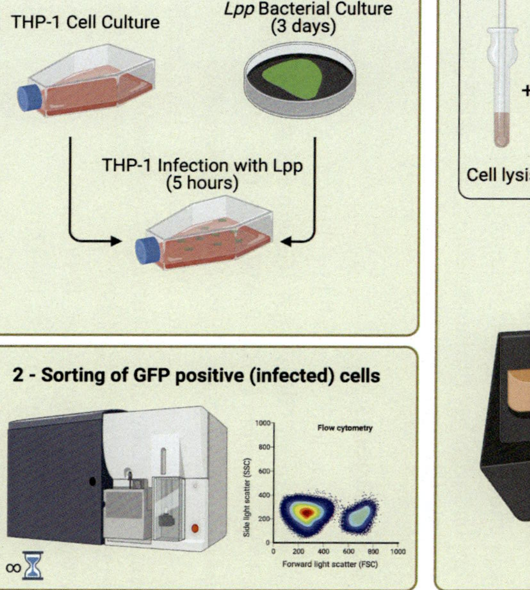

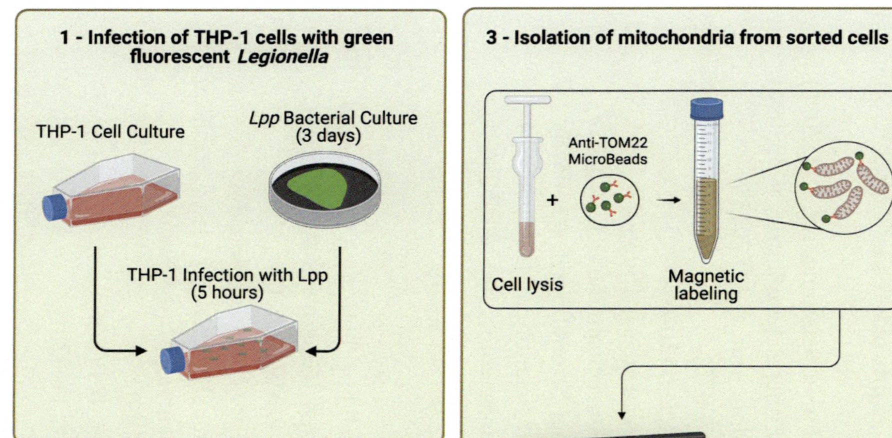

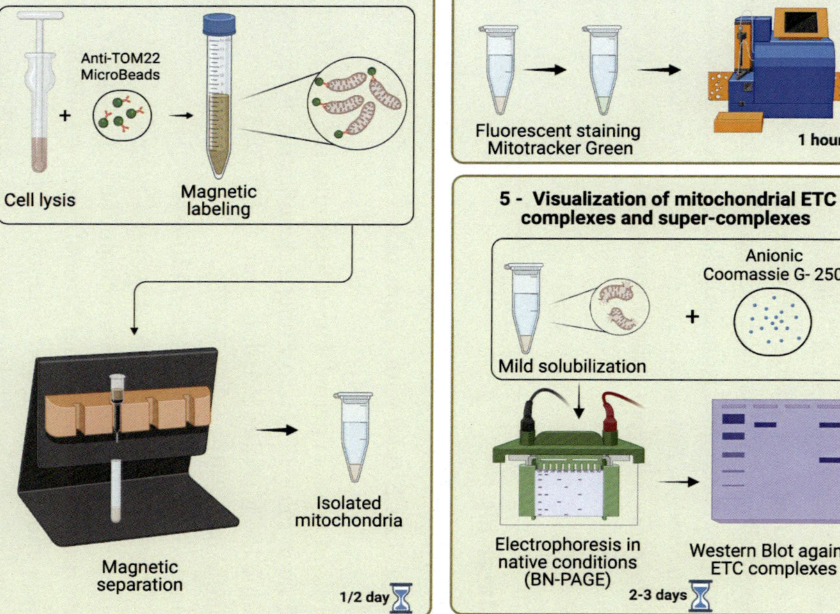

Fig. 1 Workflow for the visualization of mitochondrial ETC complexes and super-complexes. The different steps of the protocol are depicted and are numbered as appearing in the text. Approximate timing of each step is also indicated. ∞ = variable time, depending on the number of cells to sort and efficiency of bacterial infection.

THP-1 cells and GFP-expressing *L. pneumophila*, but the method can be adapted to analyze the configuration of ETC complexes and SCs in different mammalian cells infected with different intracellular bacteria expressing a fluorescent protein. Therefore, with some adaptations, this protocol should allow the study of ETC configuration during infection of mammalian cells by intracellular pathogenic bacteria, such as *Salmonella enterica*, *Mycobacterium tuberculosis*, *Listeria monocytogenes*, or *Chlamydia trachomatis*, which might shed light on cell type- and/or pathogen-specific ETC adaptations during infection.

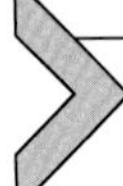

2. Materials

2.1 Disposables

- Polystyrene FACS tube with 35 μm nylon mesh filter (Corning, ref. #352235)
- 50 mL Falcon tubes (Thermo Fisher Scientific, ref. # 352098)
- 15 mL Falcon tubes (Thermo Fisher Scientific, ref. # 352096)
- Inoculating loops (Fisher Scientific, ref. # 12870155)
- T75 culture flask (SPL, ref. # 70075)
- LS columns (Miltenyi Biotec, ref. #130-042-401)
- iBlot 2 Transfer Stacks (Invitrogen, ref. # IB23001)
- PVDF membranes (Thermo Fisher Scientific, ref. # 88518)
- 10 cm Petri dish (Thermo Fisher Scientific, ref. #150318)
- 2.0 mL Eppendorf tubes (Fisher Scientific, ref. #10663981)
- 1.5 mL Eppendorf tubes (Fisher Scientific, ref. #10509691)

2.2 Cells and reagents

- Human monocytic leukemia cell line THP-1 (ATCC, ref. # TIB-202)
- RPMI 1640 medium GlutaMAX (Gibco, ref. # 61870036)
- 10% fetal bovine serum (FBS, BioWest, ref. # S1400)
- *Legionella pneumophila* strain Paris (*Lpp*) (see Note 1)
- Buffered Charcoal Yeast Extract (BYCE) agar plates (see Note 2)
- Plasmocin (InvivoGen, ref. # ant-mpp) (see Note 3)
- L-Cysteine (Sigma, ref. # 168149)
- Iron Pyrophosphate (Sigma, ref. # P6526)
- Chloramphenicol (Sigma, ref. # C0378)
- Phosphate Buffered Saline (PBS, Gibco, ref. # 10010023)

- Human Mitochondria Isolation Kit (Miltenyi Biotec, ref. # 130-094-532)
- Protease inhibitors (Roche-Sigma, ref. # COEDTAF-RO)
- Bovine Serum Albumin (BSA, Sigma, ref. # 05470)
- MitoTracker Green (Invitrogen, ref. # M7514)
- NativePAGE Sample Prep Kit (Life Technologies, ref. # BN2008)
- 20× NativePAGE running buffer (Thermo Fisher Scientific, ref. #BN2001)
- Coomassie Brilliant Blue G-250 (Serva, ref. # 17524.01)
- Bradford reagent (Thermo Fisher Scientific, ref. # J61522.AP)
- 8% acetic acid (Merck, ref. #1.09951)
- 100% methanol (Merck, ref. #322415)
- NativePAGE Bis-Tris, 3–12% gradient gel (Thermo Fisher Scientific, ref. # BN1001BOX)
- Intercept Blocking Buffer, PBS (LI-COR Biotechnology, ref. # 927-70001)
- Total OXPHOS Blue native WB Antibody Cocktail (Abcam, ref. # ab110412)
- IRDye 680RD Goat anti-Mouse (LI-COR Biotechnology, ref. # 926-68070)
- Bovine heart mitochondria (Abcam, ref. # ab110338)
- NativeMark Unstained Protein Standard (Thermo Fisher Scientific, ref. # LC0725)
- Chameleon Duo (LI-COR Biotechnology, ref. # 928-60000)

2.3 Equipment

- Spectrophotometer (Eppendorf BioPhotometer Plus)
- Benchtop cell counter, such as Countess automated Cell Counter (Invitrogen)
- Cell sorter, here we used the S3 cell sorter (Bio-Rad)
- Flow cytometer, here we used the MacsQuant VYB cytometer (Miltenyi Biotec)
- Humidified 37 °C incubator maintaining a 5% CO_2 atmosphere
- Swinging bucket centrifuge, Sorvall ST16R (Thermo Fisher Scientific)
- 1 mL Dounce glass homogenizer ×2 (Thermo Fisher Scientific)
- Tube roller mixer, such as the Cole-Parmer Stuart (Thermo Fisher Scientific)
- Magnetic separation Unit, QuadroMACS Separator (Miltenyi Biotec)

- MACS MultiStand (Miltenyi Biotec)
- iBlot 2 Gel Transfer Device (Invitrogen)
- Mini-PROTEAN Tetra Vertical Electrophoresis Cell (Bio-Rad)
- Odyssey CLx blot scanner (LI-COR Biotechnology)

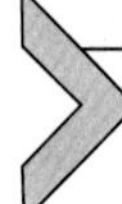

3. Methods

3.1 Cell culture, preparation of bacteria and infection

3.1.1 THP-1 cells

1. THP-1 cells are grown and maintained at 5% CO_2 and 37 °C in RPMI 1640 medium GlutaMAX (Life Technologies) supplemented with 10% FBS (see Note 3) in T75 culture flasks.

3.1.2 Bacteria

1. *Legionella pneumophila* strain Paris (*Lpp*) containing the pNT28 plasmid (see Note 1) constitutively expressing Green Fluorescent Protein (GFP) are grown on BYCE agar medium.

3.1.3 Infection

1. Retrieve *L. pneumophila* grown on BYCE for 3 days with a loop and resuspend in 1 mL of Phosphate Buffered Saline (PBS).
2. Measure the absorbance (Optical Density, OD) at 600 nm with a spectrophotometer and blank using a 1 mL PBS cuvette.
3. Adjust OD at $OD_{600} = 2.5$, which corresponds to 2.2×10^9 bacteria/mL (see Note 4).
4. Carry out further dilutions in RPMI 1640 medium in order to achieve an optimal bacterial concentration for the desired multiplicity of infection (MOI = 50 for THP-1 cells).
5. Wash THP-1 cells to remove the antibiotics present in the growth medium and resuspended in fresh RPMI without FBS.
6. Count THP-1 cells, here we used an automated cell counter.
7. Infect THP-1 cells by adding *L. pneumophila* at MOI = 50 in a T75 culture flask and incubated at 37 °C and 5% CO_2 for 5 h.
8. In parallel, a flask with the same number of non-infected THP-1 cells should be concomitantly incubated at 37 °C and 5% CO_2 for 5 h.

3.2 Fluorescence-activated cell (FACS) sorting of infected cells (see Note 5)

1. After infection, retrieve cells in a 50 mL Falcon tube completed with PBS and spun at 400*g* for 10 min at room temperature (RT). As THP-1 cells were not differentiated using phorbol 12-myristate 13-acetate (PMA) or other inducers, cells should not have attached to the T75 culture flask, being easy to retrieve them by flushing and pipetting.
2. Resuspend cells in FACS buffer composed of PBS with 2% fetal calf serum (FCS) and 2 mM EDTA with a cell concentration adjusted to $\sim 15 \times 10^6$ cells/mL and pass through a 35 μm nylon mesh filter into a polystyrene FACS tube to avoid cells aggregates.
3. Perform Flow Cytometry analysis directly after, starting with the non-infected cells to set up the gates and parameters (Fig. 2A and B).
4. Once the set-up is completed, cell sorting of infected cells can be performed using pre-set parameters (Fig. 2C).
5. Upon sorting, count cells and pellet them by centrifugation at 400*g* for 10 min. Cells can thereafter directly be used for downstream analysis or been directly put onto dry ice to flash-freeze and be kept at −80 °C for later use.

3.3 Isolation of mitochondria from sorted cells

1. Perform mitochondria isolation with the Miltenyi Biotec human Mitochondria Isolation Kit, following the corresponding protocol with few optimizations (see Note 6).
2. Thaw pellet of sorted THP-1 cells, non-infected or infected with *L. pneumophila*, from Section 3.2, step 5.
3. Depending on the infection rate and sorting efficiency, pool up to 10 tubes containing cells from the same condition in 50 mL Falcon tubes.
4. Harvest cells with a 400*g* centrifugation in a pre-cooled centrifuge, for 10 min at 4 °C.
5. Resuspend cells in ice-cold PBS with an appropriate volume approximately corresponding to 10 mL/10^7 cells according to the cell count obtained before freezing.
6. Verify again the cell number after cell resuspension with an automated cell counter.

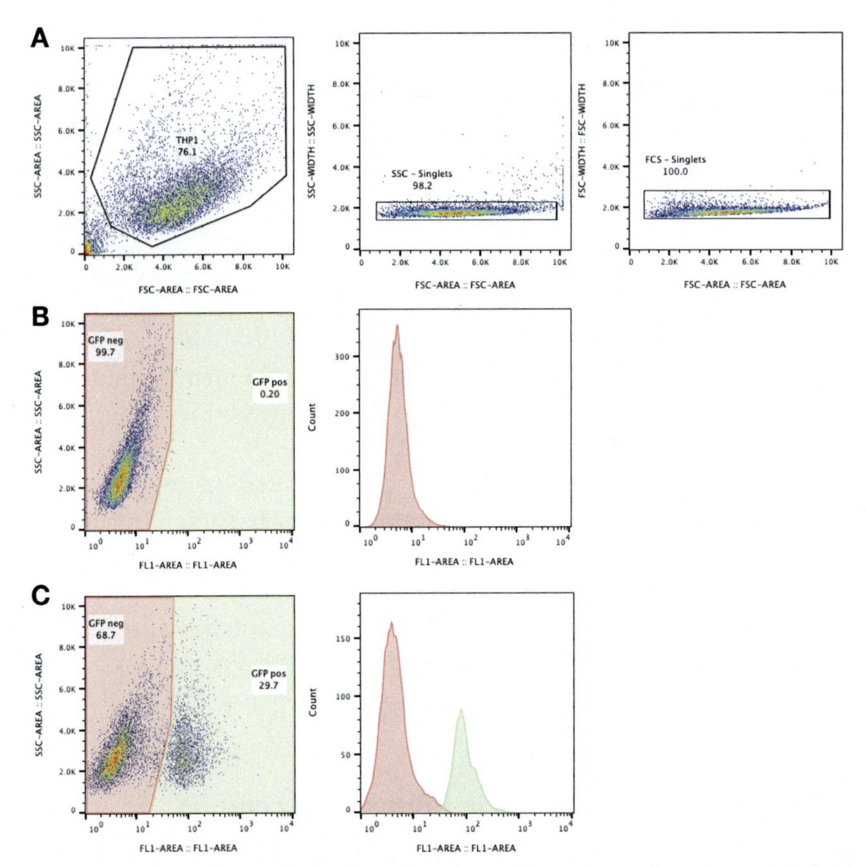

Fig. 2 Sorting of THP-1 cells infected with GFP-expressing *Legionella pneumophila* (*Lpp*). Using SSC (Side Scatter) and FSC (Forward Scatter) parameters, THP-1 cells are gated without debris. SSC-W (Width)/FSC-A (Area) and FSC-W/FSC-A allows to gate out doublets (A). Non-infected sample allows to preset the gates in order to isolate GFP-pos(itive) populations in green and GFP-neg(ative) populations in red (B). Analysis of *Lpp* WT infected THP-1 cells shows a GFP-neg cell population of 68.7% and GFP-pos cell population of 29.7% (C). The GFP-pos gate here is used for enrichment. Data was acquired with a Bio-Rad S3 cell sorter and analyzed with the FlowJo v.10.8 software.

7. Centrifuge at 400*g* for 10 min at 4 °C.
8. Discard supernatant and resuspend cells in an accurate volume of ice-cold Lysis Buffer with a volume corresponding to 1 mL of buffer for 10^7 cells.
9. Lyse cells with a Dounce homogenizer (see Note 7).
10. Transfer the lysate to a 15 mL Falcon tube.
11. Conduct magnetic labeling by adding to the cells 1× ice-cold Separation Buffer and anti-TOM22 MicroBeads to magnetically label

mitochondria (provided with the kit) with a volume corresponding to 9 mL and 50 μL, respectively, for 10^7 cells.

12. After manual homogenization by gently tilting the tube back and forth, incubate the lysate for 1 h at 4 °C under gentle shaking with a tube roller mixer.
13. Perform magnetic separation of human mitochondria labeled with anti-TOM22 magnetic beads by placing an LS column equipped with a 30 μm pre-separation filter on a magnetic separation unit.
14. Rinse LS column with 3 mL of 1× Separation Buffer.
15. Apply the lysate onto the column (see Note 8).
16. After letting the lysate completely run through, wash the column 3 times with 3 mL of 1× Separation Buffer, carefully verifying before each wash that the reservoir was emptied before starting the next wash.
17. Remove the column from the magnetic separator and placed on a 15 mL Falcon tube at safe distance from the separation unit to break the strong magnetic field.
18. Add 1.5 mL of 1× Separation Buffer.
19. Flush-out labeled mitochondria with the plunger.
20. Centrifuge isolated mitochondria at 13,000*g* for 2 min at 4 °C in a 2 mL Eppendorf tube.
21. Discard supernatant.
22. If mitochondria are not immediately used for downstream analysis, the pellet can be stored at −80 °C (Jha, Wang, & Auwerx, 2016). Otherwise, the pellet is resuspended in 150 μL of Storage Buffer provided with the isolation kit.
23. Use Bradford method to measure mitochondrial protein content concentration using bovine serum albumin (BSA) as a standard.
24. Retrieve 10 μL of mitochondrial suspension for purity assessment (Section 3.4).
25. Aliquot the desired amount of mitochondrial protein into a 1.5 mL Eppendorf tube and centrifuge at 7000*g* for 10 min at 4 °C (see Note 9).
26. After discarding the supernatant, keep the pellet on ice to proceed to Blue-Native Polyacrylamide Gel Electrophoresis (BN-PAGE).

3.4 Assessment of the integrity and purity of isolated mitochondria using a mitochondrial dye and flow cytometry

1. Resuspend a sample of the positive control, bovine heart mitochondria, and incubate in a total volume of 100 μL of Storage Buffer for staining

with MitoTracker Green (MTG, Invitrogen) at 100 nM for 30 min on ice (see Note 10).

2. A sample of isolated mitochondria non-labeled with MTG should be prepared in the same conditions as the labeled sample in step 1.
3. Treat 10 µL of mitochondria extracted from the THP-1 cells (aliquot from Section 3.3, step 25) as described in steps 1 and 2.
4. Acquire the positive control sample (bovine heart mitochondria) in the cytometer and create the gates to analyze mitochondrial populations in the extracts from infected and non-infected THP-1 cells (Fig. 3A).
5. Use the gate defined with the positive control to assess the purity and integrity of infected and non-infected THP-1 cells (Fig. 3B, see Note 11).

3.5 Visualization of mitochondrial ETC complexes and super-complexes by BN-PAGE and NIR Western blot (see Note 12)

3.5.1 *Mitochondria solubilization*

1. Heat the 5% Digitonin stock (from the NativePAGE Sample Prep Kit) at 95 °C for 3 min; afterward keep on ice until use.
2. Prepare the Sample Buffer cocktail containing digitonin, a mild detergent, following the amounts and composition given in Table 1 (see Note 13).
3. Resuspend the mitochondrial pellet that was conserved on ice after mitochondria isolation, in the appropriate volume of the Sample Buffer cocktail.
4. The required volume is calculated depending on the mitochondrial protein content concentration that was measured using the Bradford assay (see Note 14).
5. Solubilize the pellet by gently mixing with a p20 pipette tip.
6. Incubate solubilized mitochondria on ice for 20 min.
7. Centrifuge at 20,000*g* for 10 min at 4 °C.
8. Transfer 15 µL of the obtained supernatant into a new 1.5 mL Eppendorf tube.
9. Add 2 µL of Coomassie G-250 sample additive.
10. In parallel, a positive control composed of bovine heart mitochondria should be treated under the same conditions.

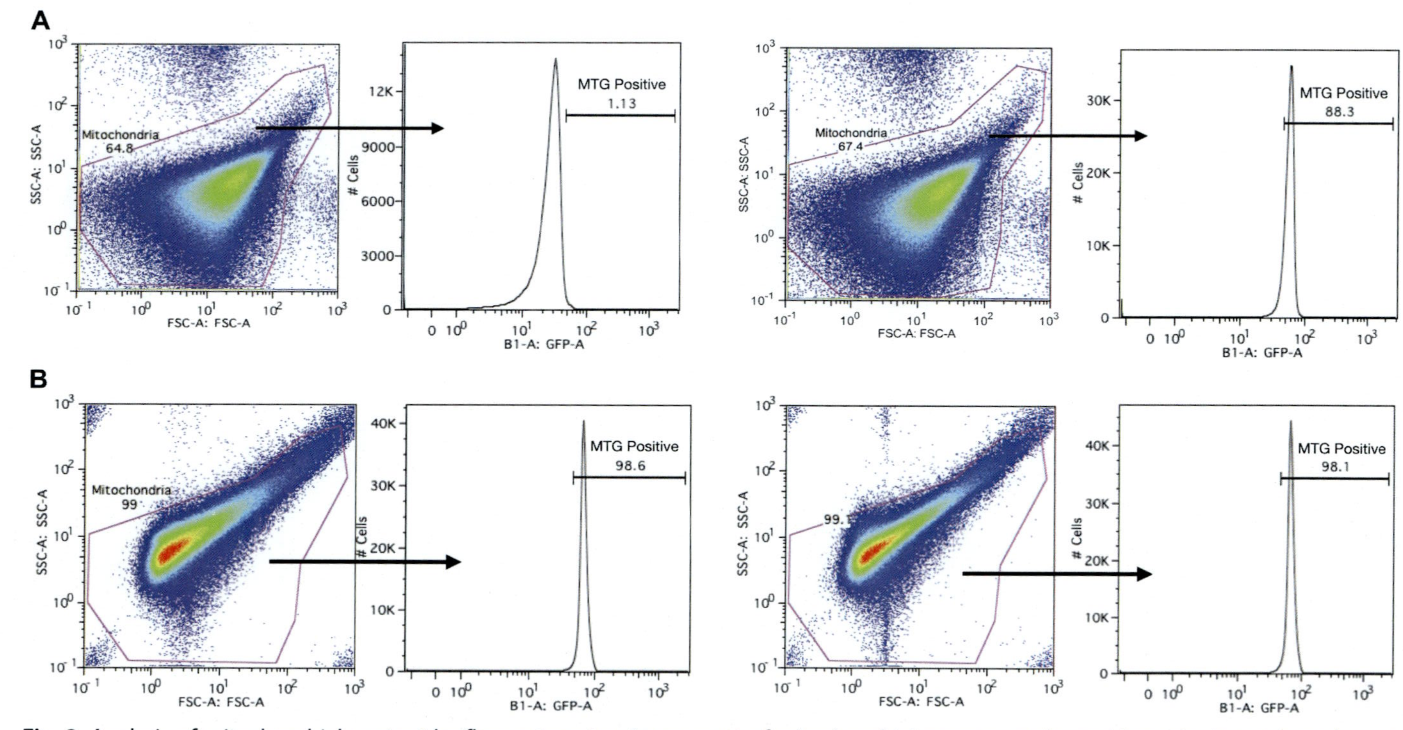

Fig. 3 Analysis of mitochondrial content by flow cytometry. Assessment of mitochondrial content, purity and integrity is conducted using MitoTracker Green (MTG), a green, fluorescent dye specific for mitochondria. (A) Bovine heart mitochondria (positive control) is used to preset the gate to analyze mitochondrial population using unlabeled mitochondria (left) and MTG-labeled mitochondria (right). (B) THP-1 cells-extracted mitochondria labeled with MTG. Mitochondria from non-infected THP-1 cells (left) and *Lpp*-infected THP-1 cells (right). Data were acquired with a MacsQuant flow cytometer and analyzed with the FlowJo v.887 software.

Table 1 Sample buffer cocktail preparation with the NativePAGE Sample Prep Kit (Thermo Fisher Scientific).

Protein	Digitonin/protein ratio (g/g)	4× Sample buffer (μL)	Digitonin (μL)	Water (μL)	Final volume (μL)
50 μg	8	5	8	7	20
50 μg	4	5	4	11	20

Example: to prepare 20 μL of sample buffer cocktail for 50 μg of protein and a Digitonin/Protein ratio of 8 (g/g), mix 5 μL of 4× Sample Buffer, 8 μL of 5% Digitonin and 7 μL of water.

3.5.2 BN-PAGE (see Note 15)

1. Prepare 700 mL of NativePAGE Anode Buffer by mixing 35 mL of 20× NativePAGE running buffer with 665 mL of distilled water.
2. Prepare the Dark Blue Cathode Buffer by diluting 0.044 g of Coomassie Brilliant Blue G-250 into 220 mL of the freshly made NativePAGE anode buffer.
3. Prepare the Light Blue Cathode Buffer by adding 20 mL of Dark Blue Cathode Buffer to 180 mL of NativePAGE Anode Buffer.
4. Load samples onto a NativePAGE Bis-Tris, 3–12% gradient gel.
5. Run the gel it in two steps (see Note 16).

3.5.3 NIR Western blot

1. Perform dry blotting of the gel using the iBlot 2 Gel Transfer Device (Invitrogen) and iBlot 2 Transfer Stacks system with PVDF membranes.
2. Blot the gel into the PVDF membrane at 20 V for 8 min.
3. To fix the proteins, wash the membrane with 8% acetic acid.
4. Wash the membrane twice with water, each time for 5 min and under gentle shaking.
5. Remove Coomassie Blue by incubating the gel three times with 100% methanol for 5 min.
6. Incubate three times in water (5 min).
7. Block the membrane for 30 min with Intercept Blocking Buffer-PBS at RT under shaking.
8. Wash with water twice for 5 min.
9. Incubate membranes overnight at 4 °C under shaking with a cocktail of primary antibodies against each mitochondrial electron chain complex using a 1/250 dilution of a Total OXPHOS Blue native WB Antibody Cocktail as listed in Table 2.

Table 2 Total OXPHOS Blue native Antibody Cocktail content and working concentrations.

Complex	Type	Clone	Target	#Ref	Working concentration
COMPLEX I	Mouse monoclonal	20C11B11B11	NDUFA9	ab14713	2 μg/mL
COMPLEX II	Mouse monoclonal	2E3GC12FB2AE2	SDHA	ab14715	0.1 μg/mL
COMPLEX III	Mouse monoclonal	13G12AF12BB11	UQCRC2	ab14745	1 μg/mL
COMPLEX IV	Mouse monoclonal	20E8C12	COX4I1	ab14744	1 μg/mL
COMPLEX V	Mouse monoclonal	15H4C4	ATP5A	ab14748	1 μg/mL

10. Next day, wash the membranes twice with PBS-Tween (0.5%, PBST).
11. Wash twice with water, each time for 5 min with gentle shaking.
12. Incubate membrane with an anti-mouse secondary antibody (1/10,000, IRDye 680RD Goat anti-Mouse) for 45 min at RT under shaking.
13. Wash two times with PBST.
14. Wash two times with water, each 5 min at RT under gentle shaking.
15. Scan the membrane using an Odyssey CLx blot scanner.

3.6 Interpretation of the results

The protocol presented here combines existing methods with the aim of visualizing mitochondrial complexes and SCs of non-infected or *L. pneumophila*-infected THP-1 cells (Fig. 4). BN-PAGE analysis allows the visualization of ETC complexes and SCs from infected human cells offering a good resolution. To help in the identification of each ETC complex and SC, we used native protein standards available for native PAGE (NativeMark Unstained Protein Standard, Thermo Fisher Scientific). As all native protein standards are unstained, they are invisible after blotting. We hence stained the gel with a Coomassie staining reagent to reveal the molecular marker bands. Expected molecular weights for ETC complexes and SCs were defined in a previous report where BN-PAGE was conducted in human cells (Greggio et al., 2017). To help identification in the NIR WB, we also added a protein standard ladder visible in NIR (Chameleon Duo, LI-COR Biotechnology).

An example of the results of the protocol presented here is summarized in Fig. 4. Samples include the positive control (bovine heart mitochondria) and mitochondria isolated from non-infected and *L. pneumophila*-infected THP-1 cells as controls (see Note 17). The native gel shows the protein ladder and the positive control stained with a Coomassie staining reagent. Blotting of the membrane with anti-ETC antibodies revealed ETC complexes and SCs by using NIR WB. A gray scale representation of the NIR WB sometimes allows an easier identification of certain complexes, such as Complex II (Fig. 4).

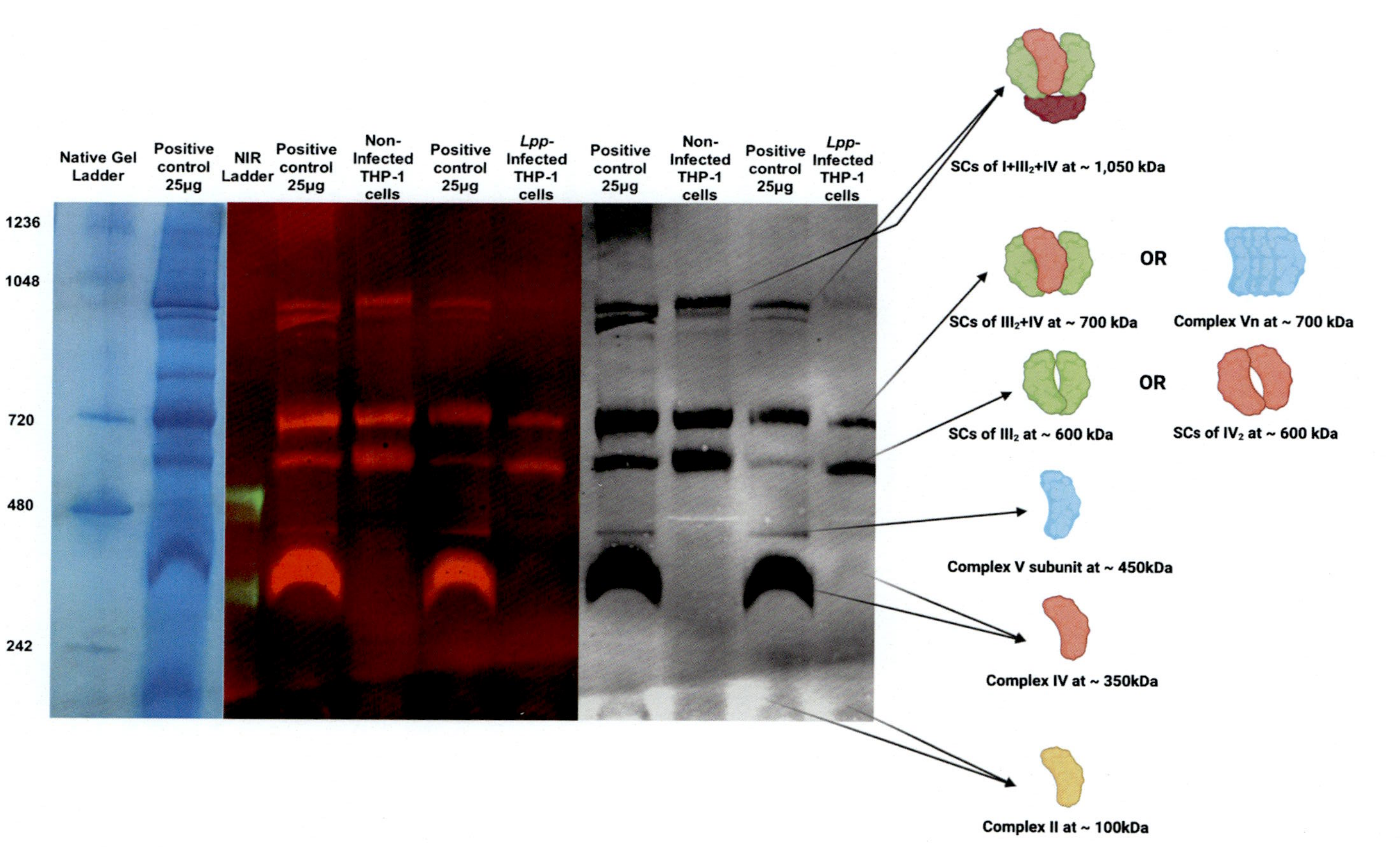

Fig. 4 See figure legend on next page.

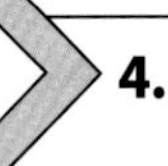

4. Notes

1. *Legionella pneumophila* strain Paris transformed with the pNT28 plasmid constitutively expressing GFP can be sent upon request, after requiring consent from the lab that originally constructed it (Tiaden et al., 2007).
2. Prepare N-(2-acetamido)-2-amino-ethanesulfonic acid (ACES)-buffered charcoal-yeast (BCYE) extract agar plate as follows: Weigh 37.5 g of N-(2-acetamido)-2-amino-ethanesulfonic acid (ACES)-buffered charcoal-yeast (BCYE) extract agar (pH 6.9, KOH) and transfer to a 1500 mL glass bottle. Add MilliQ sterile water to a volume of 1000 mL followed by autoclave for 25 min at 121 °C. Cool the agar medium to 55 °C in a water bath and supplement with 0.250 g/L of cysteine, 200 g/L pyrophosphate and 5 μg/mL of chloramphenicol. Transfer 20 mL agar medium to a 100 mm sterile Petri dish under the biosafety cabinet. Leave plate about 30 min at room temperature to solidify. Store the agar plates in a plastic bag at 4 °C until use.
3. Plasmocin, a mycoplasma prophylactic, was also added to the medium.
4. The equivalence of $OD_{600} = 2.5 = 2.2 \times 10^9$ bacteria/mL was determined using our spectrophotometer; thus it might be necessary to empirically determine the equivalence of OD_{600} and bacterial concentration if another spectrophotometer or other bacteria are used.
5. As the infection rate of THP-1 cells with *L. pneumophila* is around ~20–30%, it is necessary to avoid background due to uninfected cells during downstream analysis allowing more precise assessment of the physiology of infected cells. To obtain a population of THP-1 cells

Fig. 4 Visualization of mitochondrial complexes and SCs from non-infected and *L. pneumophila*-infected THP-1 cells by BN-PAGE and NIR Western Blot. Mitochondria were loaded on a 3–12% Bis-Tris native Gel which was then stained with Coomassie Reagent to reveal protein bands after electrophoresis at 150 V for 30 min and at 250 V for 150 min. 25 and 120 μg of protein were used for the control and the THP-1 cells samples, respectively. Left, native gel showing the protein ladder and the positive control stained with a Coomassie staining reagent. In the middle, picture of the membrane after blotting with anti-ETC antibodies and revealed using NIR WB. Samples include the positive control (bovine heart mitochondria) and mitochondria isolated from non-infected and *L. pneumophila*-infected THP-1 cells as controls. On the right is a gray scale representation of the NIR WB, which sometimes allows an easier identification of certain complexes, such as Complex II (bottom). NI: Non-infected THP-1 cells; *Lpp* WT: THP-1 cells infected with *L. pneumophila* strain Paris. Ladder values are given in kDa.

solely composed of infected cells, FACS sorting was performed. For infection we thus used a *L. pneumophila* strain constitutively expressing GFP to selectively separate GFP-positive infected cells from GFP-negative non-infected THP-1 cells.

6. The Lysis and Separation Buffer, provided with the kit, were prepared as follows: upon arrival of the kit, the Separation Buffer was incubated at 37 °C until disappearance of crystals and directly aliquoted and stored at −20 °C. Aliquots were taken out only prior to the experiment and diluted 10× with pre-cooled distilled water. On the day of the experiment, the Lysis Buffer was supplemented with protease inhibitors (Roche) to reduce the activity of proteases and phospholipases and subsequent damage of mitochondria. These two, along with PBS, were kept on ice until use.
7. The optimal stroke number was determined at ~55 strokes by monitoring cell lysis with trypan blue staining according to the method detailed previously (Carvalho et al., 2020).
8. The lysate should be applied stepwise due to the volume limitation of the column reservoir.
9. As the separation buffer contains a stabilizing agent, it is important to wash the isolated mitochondria before measuring the protein content.
10. Quality control and purity of isolated mitochondria is performed by staining the organelles with MitoTracker. Incorporation of MitoTracker dye by mitochondria is very specific for this organelle and needs an intact $\Delta\Psi_m$ in the organelle to be incorporated.
11. Our protocol allowed to successfully isolate mitochondria from THP-1 cells with a purity of ~99%. It should be noted that mitochondria from infected and non-infected cells did not seem to differ in terms of size as shown by the Forward Scatter and Side Scatter dot plots (FSC/SSC, Fig. 3B), indicating that this isolation method is very efficient for the isolation of mitochondria from THP-1 cells.
12. To visualize mitochondrial complexes and SCs with high resolution, BN-PAGE is conducted as detailed previously (Jha et al., 2016; Schägger & Pfeiffer, 2000), with some optimizations.
13. The digitonin/protein ratio was optimized to obtain the best gel resolution while taking the protein concentration into account.
14. A considerable amount of protein (~100 μg) is required in order to be able to observe SCs.
15. For BN-PAGE, all buffers need to be freshly prepared before running the gel and used immediately.

16. The outside chamber of the electrophoresis system is filled with the NativePAGE Anode Buffer during the entire electrophoresis time while the inner chamber is filled with the Dark Blue Cathode Buffer for a first run at 150 V for 30 min, then with the Light Blue Cathode buffer for a second run at 250 V for 150 min.

17. It is important to note that although THP-1 cells constitute a good model for human macrophages, cancer cells usually have a modified metabolism, therefore the obtained results might vary when primary cells are used. Indeed, we observed that Complex IV was completely absent in the mitochondria isolated from THP-1 cells, although it was present in the positive control (bovine heart mitochondria). Indeed, Greggio et al. opted to use an antibody targeting the subunit 1 of Complex IV as opposed to the antibody targeting subunit 4 of Complex IV (COX4I1, UniProt number P13073) present in the total OXPHOS antibody cocktail that we used (Greggio et al., 2017). Studies have shown that several isoforms of this subunit 4 exist and that these are regulated to optimize respiration in hypoxic cells (Fukuda et al., 2007). As THP-1 cells are originally cancer cells, it is possible that the isoform targeted by the antibody used might not be present here, hence showing a lack of signal. Therefore, the right choice of antibodies is of great importance. Using an antibody cocktail against ETC complexes as presented here allows a rapid identification and comparison among several samples of all complexes and SCs. However, to unequivocally identify the complexes participating in the composition of the different observed SCs, one should consider analyzing key samples using single antibodies for each complex during immunoblotting rather than an antibody cocktail, as well as performing sequential blots using one antibody or two antibodies at a time if they are raised in different species.

5. Concluding remarks

As the ultrastructure of mitochondria is pivotal for their respiratory activity (Cogliati et al., 2016) and mitochondrial respiration is altered when macrophages are infected by intracellular bacteria (Escoll & Buchrieser, 2019), investigating the composition of mitochondrial ETC complexes and SCs in these immune cells might be key to understand how mitochondrial metabolism is reprogrammed during infection.

By sorting infected macrophages before mitochondrial isolation, this protocol reduces the background signal caused by a high number of non-infected cells that would be in the sample (up to 70% of the cells are often not infected in the case of *L. pneumophila* infection). This background signal significantly reduces the possibility of identifying possible changes in the composition of complexes and SCs when infected and non-infected samples are compared. Therefore, sorting infected cells before isolating mitochondria is a key part of the protocol when using intracellular bacteria with infection rates below 80%.

In case primary cells are used (such as murine bone marrow-derived macrophages, BMDM, or human monocyte-derived macrophages, hMDMs) and they are challenged with GFP-labeled bacteria at a high MOI, macrophages may show infection rates of over 80% at 30 min after infection, which would make the sorting step unnecessary. It also has been demonstrated that the analysis of ETC SCs can be performed in total cell homogenates without the need of isolating mitochondria (Garaude et al., 2016), which might be a possibility if a limited number of cells are available for instance, when using primary cells. We have also used NIR WB to visualize BN-PAGE results. NIR-fluorescent blotting has proven to be more sensitive than chemiluminescent blotting because the low background fluorescence of the membrane yields a high signal-to-noise ratio, and it also has a broad linear dynamic range for quantitative measurement of WB signals (Weldon, Ambroz, Schutz-Geschwender, & Olive, 2008). Moreover, NIR WB also allows using a two-color simultaneous detection of two antigens on the same blot, which might be useful to differentiate the composition of SCs that are of similar sizes, such as SC of III_2 or IV_2 (see Fig. 4) by using e.g. mouse and rabbit antibodies coupled to secondary antibodies showing different NIR fluorescence. As explained in Note 17, if possible, blots with single antibodies and sequential blots should be used to unequivocally identify the exact composition of each SC in the WB.

Taken together, the method presented here allows to visualize the composition of complexes and SCs from mitochondria here isolated from *L. pneumophila*-infected THP-1 cells, a macrophage-like human cell line. The direct visualization of the native organization of these cellular structures and their possible modifications during infection, provides a complementary research approach to phenotypic observations of mitochondrial behavior. The combination of methods presented here can be adapted to investigate infection of macrophages by other intracellular bacteria, or to study other types of host cells, as long as the pathogen used for intracellular infection expresses a fluorescent protein allowing the sorting of infected cells prior to mitochondrial isolation.

Acknowledgments

We acknowledge all members of the BBI Unit for fruitful discussions we had about the project. This research was funded by the Institut Pasteur, the DARRI—Institut Carnot—Microbe et santé program (grant number INNOV-SP10-19) to P.E.; the Agence National de Recherche (grant number ANR-10-LABX-62-IBEID to C.B. and ANR-21-CE15-0038-01 to P.E.) and the Fondation de la Recherché Médicale (FRM) (grant number EQU201903007847) to C.B. M.D. was supported by the Ecole Doctorale FIRE—"Programme Bettencourt." D.S. was supported by the École Doctorale "ED515: Complexité du vivant."

Author contributions

M.D. performed the experiments and analyzed the data. D.S. provided important technical help for FACS sorting experiments. P.E. conceived and designed the project. P.E. and C.B. provided funding. C.B. provided critical advice and edited the manuscript. M.D. and P.E. wrote the manuscript. All authors read, revised and approved the submitted version.

Conflicts of interest

The authors declare no conflict of interest.

References

Acín-Pérez, R., Fernández-Silva, P., Peleato, M. L., Pérez-Martos, A., & Enriquez, J. A. (2008). Respiratory active mitochondrial supercomplexes. *Molecular Cell, 32*, 529–539. https://doi.org/10.1016/j.molcel.2008.10.021.

Carvalho, F., Spier, A., Chaze, T., Matondo, M., Cossart, P., & Stavru, F. (2020). Listeria monocytogenes exploits mitochondrial contact site and cristae organizing system complex subunit Mic10 to promote mitochondrial fragmentation and cellular infection. *mBio, 11*, e03171–19. https://doi.org/10.1128/mBio.03171-19.

Cogliati, S., Enriquez, J. A., & Scorrano, L. (2016). Mitochondrial cristae: Where beauty meets functionality. *Trends in Biochemical Sciences, 41*, 261–273. https://doi.org/10.1016/j.tibs.2016.01.001.

Dramé, M., Buchrieser, C., & Escoll, P. (2020). Danger-associated metabolic modifications during bacterial infection of macrophages. *International Immunology, 32*, 475–483. https://doi.org/10.1093/intimm/dxaa035.

Escoll, P., & Buchrieser, C. (2018). Metabolic reprogramming of host cells upon bacterial infection: Why shift to a Warburg-like metabolism? *The FEBS Journal, 285*, 2146–2160. https://doi.org/10.1111/febs.14446.

Escoll, P., & Buchrieser, C. (2019). Metabolic reprogramming: An innate cellular defence mechanism against intracellular bacteria? *Current Opinion in Immunology, 60*, 117–123. https://doi.org/10.1016/j.coi.2019.05.009.

Escoll, P., Platon, L., Dramé, M., Sahr, T., Schmidt, S., Rusniok, C., et al. (2021). Reverting the mode of action of the mitochondrial FOF1-ATPase by Legionella pneumophila preserves its replication niche. *eLife, 10*, e71978. https://doi.org/10.7554/eLife.71978.

Escoll, P., Song, O.-R., Viana, F., Steiner, B., Lagache, T., Olivo-Marin, J.-C., et al. (2017). Legionella pneumophila modulates mitochondrial dynamics to trigger metabolic repurposing of infected macrophages. *Cell Host & Microbe, 22*, 302–316.e7. https://doi.org/10.1016/j.chom.2017.07.020.

Fukuda, R., Zhang, H., Kim, J. W., Shimoda, L., Dang, C. V., & Semenza, G. L. (2007). HIF-1 regulates cytochrome oxidase subunits to optimize efficiency of respiration in hypoxic cells. *Cell, 129*(1), 111–122. https://doi.org/10.1016/j.cell.2007.01.047.

Garaude, J., Acín-Pérez, R., Martínez-Cano, S., Enamorado, M., Ugolini, M., Nistal-Villán, E., et al. (2016). Mitochondrial respiratory-chain adaptations in macrophages contribute to antibacterial host defense. *Nature Immunology, 17,* 1037–1045. https://doi.org/10.1038/ni.3509.

García-Rodríguez, F. J., Buchrieser, C., & Escoll, P. (2023). Legionella and mitochondria, an intriguing relationship. *International Review of Cell and Molecular Biology, 374,* 37–81. https://doi.org/10.1016/bs.ircmb.2022.10.001.

Greggio, C., Jha, P., Kulkarni, S. S., Lagarrigue, S., Broskey, N. T., Boutant, M., et al. (2017). Enhanced respiratory chain supercomplex formation in response to exercise in human skeletal muscle. *Cell Metabolism, 25,* 301–311. https://doi.org/10.1016/j.cmet.2016.11.004.

Jha, P., Wang, X., & Auwerx, J. (2016). Analysis of mitochondrial respiratory chain supercomplexes using blue native polyacrylamide gel electrophoresis (BN-PAGE). *Current Protocols in Mouse Biology, 6,* 1–14. https://doi.org/10.1002/9780470942390.mo150182.

Lenaz, G., & Genova, M. L. (2007). Kinetics of integrated electron transfer in the mitochondrial respiratory chain: Random collisions vs. solid state electron channeling. *American Journal of Physiology. Cell Physiology, 292,* C1221–C1239. https://doi.org/10.1152/ajpcell.00263.2006.

Marchi, S., Morroni, G., Pinton, P., & Galluzzi, L. (2022). Control of host mitochondria by bacterial pathogens. *Trends in Microbiology, 30,* 452–465. https://doi.org/10.1016/j.tim.2021.09.010.

Martínez-Reyes, I., & Chandel, N. S. (2020). Mitochondrial TCA cycle metabolites control physiology and disease. *Nature Communications, 11,* 102. https://doi.org/10.1038/s41467-019-13668-3.

Mitchell, G., Chen, C., & Portnoy, D. A. (2016). Strategies used by bacteria to grow in macrophages. *Microbiology Spectrum, 4.* https://doi.org/10.1128/microbiolspec.MCHD-0012-2015.

Mosser, D. M., & Edwards, J. P. (2008). Exploring the full spectrum of macrophage activation. *Nature Reviews. Immunology, 8,* 958–969. https://doi.org/10.1038/nri2448.

Nolfi-Donegan, D., Braganza, A., & Shiva, S. (2020). Mitochondrial electron transport chain: Oxidative phosphorylation, oxidant production, and methods of measurement. *Redox Biology, 37,* 101674. https://doi.org/10.1016/j.redox.2020.101674.

O'Neill, L. A. J., & Pearce, E. J. (2016). Immunometabolism governs dendritic cell and macrophage function. *The Journal of Experimental Medicine, 213,* 15–23. https://doi.org/10.1084/jem.20151570.

Schägger, H., & Pfeiffer, K. (2000). Supercomplexes in the respiratory chains of yeast and mammalian mitochondria. *The EMBO Journal, 19*(8), 1777–1783. https://doi.org/10.1093/emboj/19.8.1777.

Tiaden, A., Spirig, T., Weber, S. S., Brüggemann, H., Bosshard, R., Buchrieser, C., et al. (2007). The Legionella pneumophila response regulator LqsR promotes host cell interactions as an element of the virulence regulatory network controlled by RpoS and LetA. *Cellular Microbiology, 9,* 2903–2920. https://doi.org/10.1111/j.1462-5822.2007.01005.x.

Tiku, V., Tan, M.-W., & Dikic, I. (2020). Mitochondrial functions in infection and immunity. *Trends in Cell Biology, 30,* 263–275. https://doi.org/10.1016/j.tcb.2020.01.006.

Van den Bossche, J., O'Neill, L. A., & Menon, D. (2017). Macrophage immunometabolism: Where are we (going)? *Trends in Immunology, 38,* 395–406. https://doi.org/10.1016/j.it.2017.03.001.

Weldon, S., Ambroz, K., Schutz-Geschwender, A., & Olive, D. M. (2008). Near-infrared fluorescence detection permits accurate imaging of loading controls for Western blot analysis. *Analytical Biochemistry*, *375*, 156–158. https://doi.org/10.1016/j.ab.2007.11.035.

West, A. P., Shadel, G. S., & Ghosh, S. (2011). Mitochondria in innate immune responses. *Nature Reviews. Immunology*, *11*, 389–402. https://doi.org/10.1038/nri2975.

Zhou, R., Yazdi, A. S., Menu, P., & Tschopp, J. (2011). A role for mitochondria in NLRP3 inflammasome activation. *Nature*, *469*, 221–225. https://doi.org/10.1038/nature09663.

CHAPTER THREE

Quantitative assessment of mitochondrial membrane potential in macrophages in sepsis

Ajaz Ahmad[a], Paulraj Kanmani[a], and Guochang Hu[a,b,]*

[a]Department of Anesthesiology, University of Illinois College of Medicine, Chicago, IL, United States
[b]Department of Pharmacology & Regenerative Medicine, University of Illinois College of Medicine, Chicago, IL, United States
*Corresponding author: e-mail address: gchu@uic.edu

Contents

Abstract

Sepsis, a life-threatening condition characterized by dysregulated host response to infection, poses a significant public healthcare challenge. Excessive inflammatory responses during sepsis can lead to mitochondrial dysfunctions, resulting in organ damage. One hallmark of mitochondrial dysfunction is the reduction of mitochondrial membrane potential, which disrupts cellular metabolism, bioenergetics, and decreases the production of high-energy ATP through oxidative phosphorylation. In human sepsis, the mitochondrial membrane potential in peripheral blood monocytes has been identified as a marker of disease severity. Here, we present a detailed and widely accepted protocol for the detection of mitochondrial membrane potential using the

Methods in Cell Biology, Volume 194
ISSN 0091-679X
https://doi.org/10.1016/bs.mcb.2024.01.004

JC-1 fluorescent dye in murine bone marrow-derived macrophages and J774A.1 macrophages following stimulation with lipopolysaccharides. This protocol is routinely employed and can be easily adapted for various cell types, intact tissues, and isolated mitochondria with minimal modifications. By utilizing this technique, researchers can gain valuable insights into mitochondrial function in different experimental contexts, potentially advancing our understanding of the pathogenesis and treatment of sepsis-related mitochondrial dysfunction.

1. Introduction

Sepsis represents a paramount healthcare concern and continues to be the foremost cause of mortality among patients undergoing ICU treatment worldwide (Maslove et al., 2022; Thompson et al., 2018). The vast majority of these fatalities result from life-threatening organ dysfunction triggered by an imbalanced host response to infection (Singer et al., 2016). Sepsis is a severe and widespread inflammatory disorder that dysregulates the immune system, resulting from an unbridled response to infection (Mohanty et al., 2023). Uncontrolled inflammation throughout the body can give rise to tissue damage, multiple organ failure, and ultimate death (Poll et al., 2017, 2021).

The proper functioning of mitochondria is vital for cellular metabolism in all organs (Zhu et al., 2021). Impaired mitochondrial function can contribute to organ dysfunction and immune dysregulation in severe conditions such as sepsis (Exline & Crouser, 2008; Lelubre & Vincent, 2018; Singer, 2007). In sepsis, an overwhelming release of inflammatory mediators, including nitric oxide, and reactive oxygen/nitrogen species, can disrupt mitochondrial function (Bosmann & Ward, 2013; Galley, 2011; García et al., 2019; Jarczak et al., 2021; Mantzarlis et al., 2017). Mitochondrial dysfunction induced by sepsis disrupts the electron transport chain and further contributes to hyper-inflammation due to cellular metabolic disorders, insufficient energy production, and oxidative stress (Hotchkiss et al., 2016; Nedeva et al., 2019). This renders it a pivotal factor in the development of multiorgan failure associated with sepsis (Arulkumaran et al., 2016; Chen et al., 2011; Merz et al., 2017; Shenoy et al., 2020; Singer, 2014; Sjövall et al., 2013; Víctor et al., 2009). During sepsis, mitochondrial dysfunction can result in persistent cellular and organ damage caused by elevated levels of reactive oxygen species (ROS), reduced oxygen consumption, opening of mitochondrial permeability transition, and decreased mitochondrial membrane potential (Cimolai et al., 2015; Halestrap et al., 2002; Ricquier & Bouillaud, 2000).

The mitochondrial membrane potential serves as a crucial indicator of mitochondrial activity, as it reflects the intricate processes of electron transport and oxidative phosphorylation that drive ATP production (Zorova et al., 2018). Maintaining the normal mitochondrial transmembrane potential is vital for sustained ATP generation throughout the lifespan of cells (Bagkos et al., 2014). Inflammation can induce a significant loss of mitochondrial membrane potential through the impact of ROS on cardiolipin and components of the mitochondrial permeability transition pore, such as voltage-dependent anion channel, adenine nucleotide translocator and cyclophilin D (Bonora et al., 2022; Lopes-Pires et al., 2021; Potz et al., 2016; Yao et al., 2015). Higher mitochondrial membrane potential in monocytes has been associated with reduced organ injury (Weiss et al., 2015). The reduction of mitochondrial membrane potential is considered an initial and irreversible step toward apoptosis and cell death (Adrie et al., 2001). A critical decline in mitochondrial membrane potential results in energy depletion, leading to organ failure and mortality in septic patients (Lee & Huttermann, 2014). Thus, mitochondrial transmembrane potential serves as a marker of severity in human sepsis, with a more pronounced reduction observed in non-survivor septic patients compared to survivors (Adrie et al., 2001).

The mitochondrial membrane potential generated by proton pumps (Complexes I, III and IV) is a crucial parameter of mitochondrial function and serves as an indicator of cellular health. Assessment of the mitochondrial membrane potential in immune cells including macrophages can be used to estimate the mitochondrial bioenergetic state and the severityof sepsis. The non-toxic fluorescent probe 5,5,6,6′-tetraethylbenzimidazolylcarbocyanine iodide (JC-1) is widely used to measure mitochondrial membrane potential in a variety of cell types, intact tissues, and isolated mitochondria (Perelman et al., 2012). The JC-1 dye demonstrates the selective accumulation in mitochondria that is dependent on membrane potential. This accumulation is indicated by a green fluorescence emission at approximately 529 nm for the monomeric form of the probe. With a concentration-dependent formation of red fluorescent J-aggregates, the emission shifts to red at around 590 nm. Dye color changes reversibly from red to green as mitochondrial membrane potential decreases due to depolarization. At lower membrane potential or internal mitochondrial concentrations, the JC-1 dye exists as monomers. However, at higher membrane potentials or concentrations, the JC-1 dye monomers form red fluorescent J-aggregates within the mitochondria where they have accumulated. Importantly, the ratio of red to green fluorescence is solely dependent on the mitochondrial membrane potential and is unaffected by factors such as mitochondrial size, shape, and density, which can influence single-component fluorescence signals (Chazotte, 2011). In normal cells, JC-1 readily accumulates within the energized and negatively charged mitochondria, leading to the formation of vibrant red fluorescent J-aggregates. However, in infected or apoptotic cells, where the membrane permeability is increased and the membrane potential is lost, JC-1 exhibits reduced penetration into the mitochondria, resulting in lower levels of red fluorescent J-aggregates (Elefantova et al., 2018). By utilizing the red-green fluorescence ratio, researchers can perform comparative measurements of membrane potential, assess the percentage of mitochondrial depolarization in pathological conditions, and study the impact of sepsis on mitochondrial function (e.g., cellular stress and apoptosis) (Kanmani et al., 2023; Sivandzade et al., 2019). The primary aim of this paper is to present a practical, detailed protocol for assessing and monitoring mitochondrial membrane potential in whole cells using the JC-1 fluorescent cationic probe.

p120 Catenin (p120) belongs to the armadillo protein family and is known for its role in binding to the cytoplasmic domain of classical cadherins, facilitating the formation of adherens junctions in both endothelial and epithelial cells (Hu, 2012). Our previous research has established the pivotal significance of p120 expression in macrophages for anti-inflammatory responses during sepsis (Yang et al., 2014). Notably, depleting p120 using a specific siRNA in macrophages has been demonstrated to disrupt mitochondrial function and reduce mitochondrial membrane potential following lipopolysaccharides (LPS) stimulation (Kanmani et al., 2023). Here, we provide an illustrative example through a comprehensive data analysis, encompassing measurements of mitochondrial membrane potential in both bone marrow-derived macrophages (BMDMs) and J774A.1 macrophages treated with p120 siRNA in response to LPS stimulation.

2. Materials

2.1 Common disposables

- Dissection toolbox: Sterilized (see *Note 1*)
- Polystyrene dissection board (foam box cover) (see *Note 1*)
- 25-gauge needle, sterile (BD, #305122) (see *Note 1*)
- 15 mL centrifuge tube, sterile (Denville, #C1071p) (see *Note 1*)

- 10 mL syringe (BD, #303134) (see *Note 1*)
- 50 mL Corning tube, sterile (Denville, #C1062p) (see *Notes 1 and 2*)
- 6-well tissue culture plate (Denville, #1020A) (see *Notes 1 and 3*)
- 60 mm tissue culture dish (Denville, #T1106) (see *Notes 1 and 4*)
- 70 μm pore size cell strainer (Falcon®, #352350) (see *Note 1*)
- Ultra Cruz Syringe Filters, PVDF, 0.22 μm (see *Note 1*)
- 10 cm culture dish (Denville, #T1110) (see *Note 3*)

2.2 Cells and reagents

- BMDMs: isolated from C57BL/6 mice (see *Note 5*)
- J774A.1 cells: isolated from the ascites of an adult female mouse with reticulum cell sarcoma (ATCC, Rockville, MD) (see *Note 5*)
- Fetal Bovine Serum (FBS) (Sigma, #F2442) (see *Note 6*)
- Trypan blue solution (0.4% solution in PBS) (Fisher Scientific, #15250061) (see *Note 1*)
- DMEM (Dulbecco's modified Eagle medium) (Corning, #10-013-CV) (see *Note 5*)
- 100 U/mL penicillin-streptomycin (Fisher Scientific, #10378016) (see *Note 1*)
- RPMI 1640 medium (Corning, #10-040-CV) (see *Note 1*)
- Ca^{2+} and Mg^{2+} free phosphate buffered saline (DPBS) (Corning, #21-031-CV) (see *Note 1*)
- Mitochondrial-specific probe cyanine dye JC-1 (5,5′,6,6′-tetrachloro-1,1′,3,3′-tetraethylbenzimi-dazolylcarbocyanine iodide, Santa Cruz, SC-364116A) (see *Notes 7 and 8*)
- 70% ethanol (v/v in distilled water) (Decon Laboratories Inc. DSP-MD.43) (see *Note 1*)
- Trypsin-EDTA solution (5×, Sigma, T3924) (see *Note 9*)
- Ketamine (Fort Dodge IA, #0856-2013-01) (see *Note 1*)
- Xylazine (Ben Venue Lab, #139-236) (see *Note 1*)

2.3 Equipment

- Inverted fluorescence microscope (60 X magnifications) (see *Note 1*)
- Surgical scissors (Roboz Surgical Instrument Co., RS-5980, see *Note 1*)
- Fine forceps (Fine Science Tools, 11009-13, see *Note 1*)
- Curved forceps (Fine Science Tools, 14060-10, see *Note 1*)
- Centrifuge (Eppendorf, 5415R) (see *Note 1*)
- 37 °C CO_2 incubator (Isotemp, Fisher Scientific) (see *Note 1*)

- Cell culture hood (Sterilgard hood) (see *Note 1*)
- Hemocytometer (Hausser Scientific) (see *Note 1*)
- Purifier Logic plus Class II, Type A2 Biosafety Cabinet used for cell operations (see *Note 10*)

2.4 Software

- Fluorescence intensity was analyzed with the help of software (Image J (NIH)) (see *Note 11*)
- Prism (Graph Pad, Version 9.2.0) (see *Note 12*)

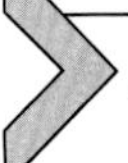

3. Method

3.1 Isolation and culture of bone marrow-derived macrophages (BMDMs)

1. Administer ketamine (100 mg/kg, i.p.) and xylazine (10 mg/kg, i.p.) to induce anesthesia in mice, followed by euthanization through cervical dislocation. These procedures were conducted in accordance with the guidelines approved by the Institutional Animal Care and Use Committee of the University of Illinois at Chicago
2. Sterilize the outer skin of the mouse by gently spraying 70% ethanol (see *Note 13*)

   Carefully peel off the skin from the leg of the mice and make an incision at the hip joint. Transfer the leg to a 60 mm dish filled with ice-cold sterile PBS
3. Excise the muscles from the femurs and tibias, and then transfer them to a 60 mm dish filled with 70% ethanol. After a 2-min immersion, transfer the bones to a new dish containing RPMI medium supplemented with antibiotics, penicillin (100 U/mL), and streptomycin (100 μg/mL) (see *Notes 5 and 14*)
4. Using sterile tweezers, grip one bone and use sterile scissors to carefully cut both ends, allowing access to the bone marrow
5. Take a 10 mL syringe and affix a 25G needle. Draw RPMI medium supplemented with antibiotics (penicillin and streptomycin) into the syringe
6. Carefully insert the 25G needle into the bone cavity and extract the bone marrow into a fresh 60 mm dish. Continuously repeat steps 6–7 until the bone turns white in color
7. To optimize cell dissociation, gently aspirate the medium containing the bone marrow using the same syringe and pass it through the needle.

For optimal results, it is crucial to strain the cell suspension through a 70 μm pore size cell strainer (see *Note 1*)

8. Transfer the cell suspension into a 15 mL polypropylene centrifuge tube, centrifuge it at 300 × *g* for 5 min at room temperature, and subsequently discard the supernatant (see *Note 15*)
9. Introduce 1 mL of distilled water into the tube to lyse the red blood cells. Perform gentle pipetting for 10 s, and promptly add PBS to reach a final volume of 15 mL
10. Centrifuge the suspension containing lysed red blood cells at 300 × *g* for 5 min at room temperature. Repeat the lysis step if any residual red blood cells are still visible in the pellet
11. Discard the supernatant and carefully resuspend the cell pellet in 20 mL of differentiation medium to obtain a homogenous single-cell suspension. To assess the number of viable cells, employ a 0.4% Trypan blue solution in conjunction with a hemocytometer chamber and an optical microscope
12. Take 10 mL of the cell suspension and plate the cells onto a 100 mm tissue culture dish. Incubate the dish at 37 °C with 5% CO_2 for 90 min to allow for the removal of stromal cells
13. Collect the non-adherent cells and plate them onto a 100 mm Petri dish, ensuring a density of 4×10^5 cells per dish. Add 10 mL of differentiation medium to each dish (see *Note 16*)
14. After a culture period of 7 days, the bone marrow precursors are considered fully differentiated into macrophages, thereby rendering them ready for subsequent experimental procedures

3.2 Culture of J774A.1 murine macrophage

1. Retrieve the glycerol stock vial from the liquid nitrogen tank and place it on ice. It is important to note that the cells should be cultured in DMEM supplemented with 10% (v/v) heat-inactivated FBS, 200 U/mL penicillin-streptomycin at a temperature of 37 °C with 5% CO_2 (see *Notes 5 and 14*)
2. Inside the hood, dilute the stock by adding a warm (37 °C) DMEM medium
3. Pellet the cells by centrifuging at 1300 × *g* for 5–10 min. Remove the supernatant and resuspend the cells in 5 mL of warm DMEM
4. For cell counting, combine 100 μL of cells with 100 μL of normal saline in a 1:1 proportion. Transfer 200 μL of this mixture to the hemocytometer and carefully count the number of cells (see *Note 16*)

5. Begin the preparation for the experiment by adding the appropriate quantity of stock to the dish. Subsequently, fill the dish with media until it reaches its maximum capacity, which will vary based on the desired cell quantity for the experiment
6. Ensure the cells are evenly distributed in the dish and proceed to place the culture dishes in an incubator set at 37 °C with 5% CO_2 to facilitate optimal growth conditions
7. Regularly monitor the color of the DMEM media. If it appears light red or yellow, promptly replace it to maintain optimal conditions for cell growth (see *Note 17*)
8. Following a period of 2–3 days, the J774A.1 cell should achieve a confluency of 90%, indicating readiness for subsequent experiments (see *Note 18*)

3.3 RNA interference and treatment

1. Plate BMDMs and J774A.1 cells on a 6-well culture plate and incubate them at 37 °C with 5% CO_2 until they reach a confluency of 50–70% (see *Note 5*)
2. Incubate both BMDMs and J774A.1 cells with serum-free medium containing scrambled (SC) or p120 small interfering RNA (siRNA) (three target-specific 20- to 25-nt siRNAs, Dharmacon) at a concentration of 50 nM, following the manufacturer's instructions
3. After 24–48 h of incubation, aspirate the medium and add a fresh complete medium. Continue incubation for another 2–3 h
4. Confirm the efficacy of p120 deletion in both BMDMs and J774A.1 macrophages by performing Western blot analysis
5. Once the cells are ready, treat them with LPS (20 ng/mL) for 4 h, followed by stimulation with ATP (5 mM) for 30 min

3.4 Measurements of mitochondrial membrane potential

1. Prepare a fresh stock solution by reconstituting the lyophilized JC-1 dye with DMSO, resulting in a 100× stock solution with a concentration of 200 μM, right before experiments (see *Note 19*)
2. Ensure thorough mixing of the solution until all the dye powder is completely dissolved, and there are no visible aggregates present in the solution (see *Note 20*)
3. For proper cell culture, use gelatin pre-coated glass coverslips in a 24-well culture plate. Maintain a maximum cell density of approximately 1×10^6 cells/cm^2, as this is crucial for obtaining optimal results (see *Note 21*)

4. The precoating agent and density required may vary depending on the type of cells being used. It is strongly recommended to follow your specific cell culture protocol for guidance (see *Note 22*)
5. Remove the current cell culture medium and replace it with a pre-warmed medium at approximately 37 °C
6. Rinse the cells twice with pre-warmed PBS, and then add fresh cell culture medium to each well
7. Add 2 μM (final concentration) of JC-1 dye immediately. Incubate the cells at 37 °C with 5% CO_2 for a minimum of 15–30 min (see *Note 23*)
8. Add a final concentration of 50 μM of carbonyl cyanide-3-chlorophenyl hydrazine as a positive control and allow the cells to incubate at 37 °C for 5 min
9. Remove the culture medium from the wells and rinse the wells with pre-warmed PBS
10. Add 100 μL to each well of a 96-well plate, 500 μL to a 24-well plate, or the corresponding amount for a Petri dish or chamber slides, based on the selected culture platform and detection methods. To prevent light exposure, cover the cell culture plate and its contents with a foil sheet
11. Immediately analyze the fluorescence of both the test cultures and the controls using the quantification methods
12. The dye produces green and red fluorescence at a standard mitochondrial membrane potential when excited at 490 nm (see *Note 24*)
13. Utilize an inverted fluorescence microscope at 60× magnification to measure fluorescence. Capture images of the same microscopic field in the red and green channels and merge the resulting images (see *Note 25*)
14. It's important to note that JC-1 is a cationic dye that can accumulate in mitochondria, exhibiting a fluorescence emission shift from green (~529 nm) to red (~590 nm)
15. A flowchart illustrates the primary procedural steps involved in measuring mitochondrial membrane potential in macrophages (Fig. 1)

3.5 Data analysis

1. In this study, we investigated the role of p120 in regulating mitochondrial membrane potential in macrophages upon ATP stimulation following LPS priming, using a JC-1 fluorescent probe.
2. Treatment with LPS and ATP resulted in a reduction in the ratio of JC-1 aggregate to monomer, indicating a decrease in mitochondrial membrane potential in both control and p120 knockdown cells (Fig. 2A and B).

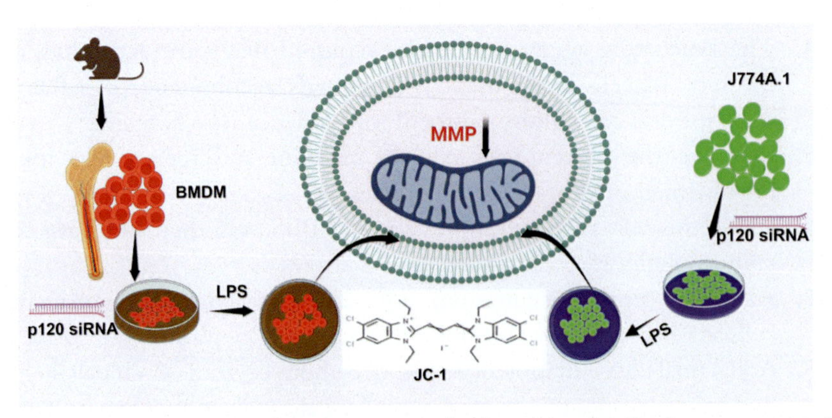

Fig. 1 Flowchart illustrating the primary procedural steps involved in the assessment of mitochondrial membrane potential in macrophages. BMDM, bone marrow-derived macrophages; JC-1, 5,5,6,6′-tetrachloro-1,1′,3,3′-tetraethylbenzimi-dazolylcarbocyanine iodide; LPS, lipopolysaccharides; MMP, mitochondrial membrane potential; p120, p120 catenin; siRNA, small interfering RNA.

3. Interestingly, p120 depletion exacerbated the LPS/ATP-induced decrease in mitochondrial membrane potential (Fig. 2A and B). These findings strongly suggest that p120 deletion leads to reduced membrane potential and mitochondrial dysfunction in macrophages during sepsis.

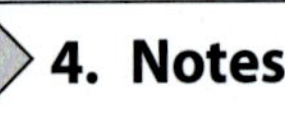

4. Notes

1. While the catalog number and provider are provided as references, it is important to note that an equivalent product can be obtained from various providers.
2. As an alternative approach, 96-well bottom plates, such as Corning® 96 Well Polystyrene Microplate #CLS3897 from Corning, can be utilized instead of the conventional method.
3. For different cell quantity requirements, larger (15 cm) or smaller (6 cm) cell culture dishes can be used accordingly.
4. When using larger (150 mm) or smaller (60 mm) cell culture dishes, the amount of 0.05% trypsin-EDTA solution should be adjusted accordingly.
5. Although BMDMs and J774.1A cells, as well as DMEM, FBS, HEPES, PBS, RPMI, and trypsin-EDTA, are classified as non-hazardous, it is still recommended to use certified personal protective equipment (PPE) when handling them.

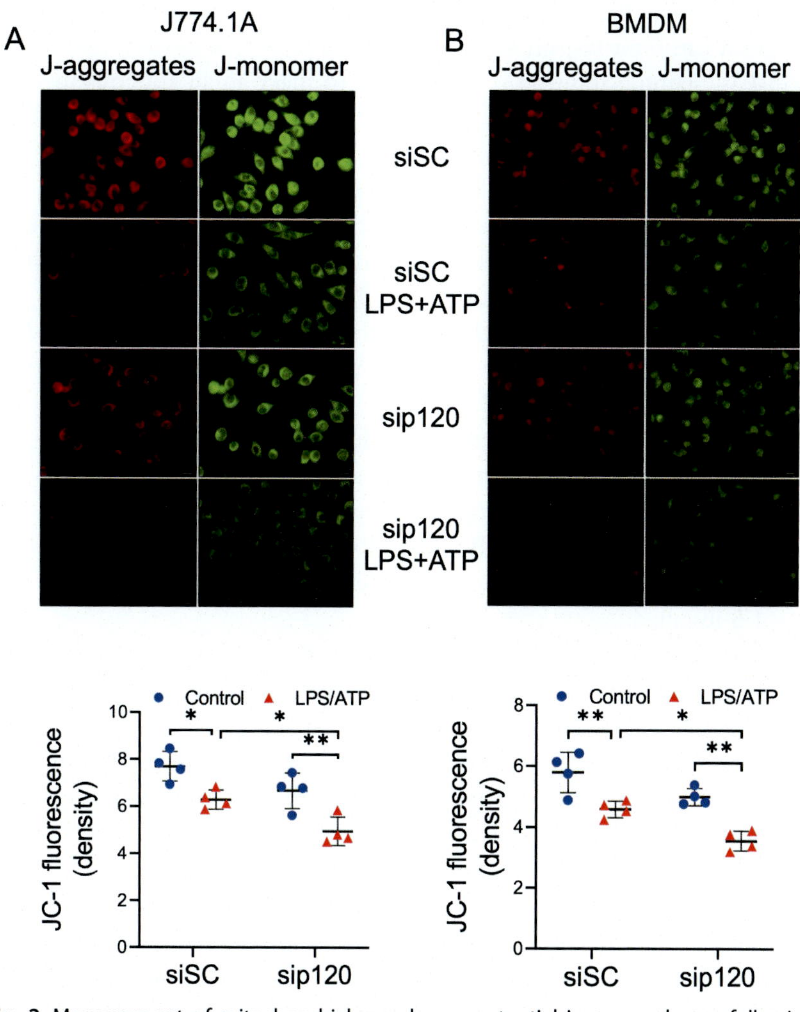

Fig. 2 Measurement of mitochondrial membrane potential in macrophages following LPS/ATP stimulation. Bone marrow-derived macrophages (BMDMs) isolated from C57BL/6J mice and J774.1A cells were cultured as detailed in Sections 3.1 and 3.2. They were then transfected with scrambled (siSC) and p120 (sip120) siRNA, as detailed in Section 3.3. After 48 h post-transfection, the cells were incubated with LPS and stimulated with ATP, as detailed in Section 3.3. Subsequently, the mitochondrial membrane potential was assessed using JC-1 (A and B), as detailed in Section 3.4. Mitochondrial membrane potential was determined using a fluorescence microscope with a magnification of ×400. (A) Mitochondrial membrane potential in J774.1A cells. (B) Mitochondrial membrane potential in BMDMs. The top panel displays the imaging of JC-1 monomer (green) and aggregate (red) fluorescence, with a scale bar of 10 μm. The bottom panel presents the histogram illustrating the distribution of JC-1 aggregate and monomer fluorescence intensity. The level of mitochondrial membrane potential was computed by obtaining the ratio of J-aggregate to J-monomer immunofluorescence intensity ($n = 4$). Data represent means ± SD. n = number of experimental replicates. $^{*}P < 0.05$, $^{**}P < 0.01$ (one-way ANOVA with Bonferroni post hoc test).

6. Removing FBS remnants will improve detachment efficiency since FBS can limit trypsin's enzymatic activity.
7. JC-1 staining solution is challenging to make in an aqueous medium due to low solubility and a tendency to form particles that are difficult to remove. Therefore, ensure that the JC-1 reagent is thoroughly dissolved before diluting it into the culture medium.
8. To avoid any interference with JC-1's light sensitivity, it is essential to conduct all staining procedures in an environment without direct exposure to intense light.
9. To optimize cell detachment efficiency, it is recommended to pre-warm trypsin-EDTA, as it exhibits the highest catalytic activity at a temperature of 37 °C.
10. All cell culture stages must be conducted within a Class II Biological Safety Cabinet to prevent contamination.
11. ImageJ is a Java-based program designed for processing and analyzing images. It is built upon NIH Image for Macintosh and is freely available in the public domain. ImageJ can be used on any computer with Java 1.5 installed.
12. Prism is preferred for analyzing scientific data and creating clear graphs. Its user-friendly interface and impressive capabilities aid in presenting scientific data with clarity.
13. Ethanol is flammable in both liquid and vapor forms, and it can cause severe eye irritation. When handling ethanol, appropriate certified personal protective equipment (PPE) should be worn, and it should be stored in flammable cabinets away from open flames.
14. Penicillin and streptomycin can cause skin irritation, allergy symptoms, breathing difficulties if inhaled, and reproductive impairment in unborn children. Therefore, it is important to wear suitable certified PPE when handling these substances.
15. Before using micro-centrifuge tubes for cell culture, autoclave them to disinfect and prevent bacterial or fungal contamination.
16. Lower cell counts should be avoided to reduce the likelihood of genetic drift caused by population bottlenecks and/or founder effects.
17. It is recommended to discard opened bottles of prepared medium containing glutamine after 4–6 weeks to avoid the risk of hazardous ammonia caused by the breakdown of glutamine.
18. Over-confluence should be avoided, as it may impair cellular development during passage due to metabolic disturbances.

19. JC-1 should be treated as a potential mutagen and handled carefully. Disposal should follow local regulations, and protective clothing and gloves should always be worn when handling this reagent.
20. Before starting, ensure that the JC-1 powder and DMSO solutions reach a temperature of 25 °C prior to use.
21. When conducting an experiment using JC-1, maintaining a temperature of 25 °C and a pH level of 7.4 throughout the experiment is crucial, as the reagents are sensitive to changes in temperature and pH.
22. The JC-1 dye staining protocol is relatively quick, with the duration of the process dependent on the number of samples being analyzed.
23. To obtain the best results when staining cells with JC-1, avoid centrifuging the staining solution, and analyze the cells promptly after washing. If the cells are not adequately stained, adding more JC-1 dye may be necessary. Additionally, avoid exposing the stained cells to intense light, as this can greatly affect the quality of the staining.
24. A low ratio of red to green signal in the untreated control cells may indicate compromised cell viability.
25. To ensure accurate analysis, promptly examine the samples using a fluorescence microscope. If immediate analysis is not possible, refrigerate the samples (not freeze) and keep them in the dark for no more than 24 h to maintain sample integrity and minimize potential errors.

5. Concluding remarks

The mitochondrial transmembrane potential acts as an indicator of severity in human sepsis, showing a more significant decrease in non-survivor septic patients compared to survivors (Adrie et al., 2001). Accurate detection of changes in mitochondrial membrane potential is crucial in determining the development of sepsis. Assessing mitochondrial membrane potential in cells can be challenging, but our study presents a simple and effective method using fluorescence microscopy and JC-1 cationic dye. This dye accumulates in the mitochondrial matrix and emits red fluorescence upon membrane potential polarization, making it an excellent mitochondrial capacity indicator. By analyzing the red-to-green fluorescence ratio exclusively determined by membrane potential, we can compare measurements and calculate the percentage of mitochondria in a population that respond to an applied stimulus. Furthermore, JC-1 is more targeted toward mitochondrial than plasma membrane potential and provides more consistent depolarization feedback than other cationic dyes. This method is highly valuable for detecting subtle cellular response differences during sepsis and other cellular events, such as apoptosis (Adrie et al., 2001; Zhang et al., 2018). Employing JC-1 and fluorescence microscopy in fixed BMDMs and J774.1A cells, our experimental protocol proves to be a reliable and effective approach for assessing mitochondrial membrane potential in sepsis.

Acknowledgments

The work was supported by an R01 grant (#HL152696, PI: G.H.) and an R01 grant (#HL104092, PI: G.H.) from the National Heart, Lung, and Blood Institute of the National Institutes of Health, and an R21 grant (#AI152249, PI: G.H.) from the National Institute of Allergy and Infectious Diseases of the National Institutes of Health.

Competing interests

The authors have no conflicts of interest to declare.

References

Adrie, C., et al. (2001). Mitochondrial membrane potential and apoptosis peripheral blood monocytes in severe human sepsis. *American Journal of Respiratory and Critical Care Medicine*, *164*, 389–395. https://doi.org/10.1164/ajrccm.164.3.2009088.

Arulkumaran, N., et al. (2016). Mitochondrial function in sepsis. *Shock*, *45*, 271–281. https://doi.org/10.1097/SHK.0000000000000463.

Bagkos, G., et al. (2014). A new model for mitochondrial membrane potential production and storage. *Medical Hypotheses*, *83*, 175–181. https://doi.org/10.1016/j.mehy.2014.05.001.

Bonora, M., et al. (2022). Molecular mechanisms and consequences of mitochondrial permeability transition. *Nature Reviews. Molecular Cell Biology*, *23*, 266–285. https://doi.org/10.1038/s41580-021-00433-y.

Bosmann, M., & Ward, P. (2013). The inflammatory response in sepsis. *Trends in Immunology*, *34*, 129–136. https://doi.org/10.1016/j.it.2012.09.004.

Chazotte, B. (2011). Labeling mitochondria with JC-1. *Cold Spring Harbor Protocols*, *2011*, pdb.prot065490. https://doi.org/10.1101/pdb.prot065490.

Chen, X. H., et al. (2011). Sepsis and immune response. *World Journal of Emergency Medicine*, *2*, 88–92. PMCID PMC4129694.

Cimolai, M. C., et al. (2015). Mitochondrial mechanisms in septic cardiomyopathy. *International Journal of Molecular Sciences*, *16*, 17763–17778. https://doi.org/10.3390/ijms160817763.

Elefantova, K., et al. (2018). Detection of the mitochondrial membrane potential by the cationic dye JC-1 in L1210 cells with massive overexpression of the plasma membrane ABCB1 drug transporter. *International Journal of Molecular Sciences*, *19*, 1985. https://doi.org/10.3390/ijms19071985.

Exline, M. C., & Crouser, E. D. (2008). Mitochondrial mechanisms of sepsis-induced organ failure. *Frontiers in Bioscience*, *13*, 5030–5041. https://doi.org/10.2741/3061.

Galley, H. F. (2011). Oxidative stress and mitochondrial dysfunction in sepsis. *British Journal of Anaesthesia*, *107*, 56–64. https://doi.org/10.1093/bja/aer093.

García, J. J. M., et al. (2019). P2X7 receptor induces mitochondrial failure in monocytes and compromises NLRP3 inflammasome activation during sepsis. *Nature Communications*, *10*, 2711. https://doi.org/10.1038/s41467-019-10626.

Halestrap, A. P., et al. (2002). The permeability transition pore complex: Another view. *Biochimie*, *84*, 153–166. https://doi.org/10.1016/s0300-9084(02)01375-5.

Hotchkiss, R., et al. (2016). Sepsis and septic shock. *Nature Reviews. Disease Primers*, *2*, 16045. https://doi.org/10.1038/nrdp.2016.45.

Hu, G. (2012). p120-Catenin: A novel regulator of innate immunity and inflammation. *Critical Reviews in Immunology*, *32*, 127–138. https://doi.org/10.1615/critrevimmunol.v32.i2.20.

Jarczak, D., et al. (2021). Sepsis—Pathophysiology and therapeutic concepts. *Frontiers in Medicine*, *8*, 628302. https://doi.org/10.3389/fmed.2021.628302.

Kanmani, P., et al. (2023). p120-Catenin suppresses NLRP3 inflammasome activation in macrophages. *American Journal of Physiology. Lung Cellular and Molecular Physiology*, *1*, L596–L608. https://doi.org/10.1152/ajplung.00328.2022.

Lee, I., & Huttermann, M. (2014). Energy crisis: The role of oxidative phosphorylation in acute inflammation and sepsis. *Biochimica et Biophysica Acta*, *1842*, 1579–1586. https://doi.org/10.1016/j.2014.05.031.

Lelubre, C., & Vincent, J. (2018). Mechanisms and treatment of organ failure in sepsis. *Nature Reviews. Nephrology*, *14*, 417–427. https://doi.org/10.1038/s41581-018-0005-7.

Lopes-Pires, M. E., et al. (2021). Clotting dysfunction in sepsis: A role for ROS and potential for therapeutic intervention. *Antioxidants (Basel)*, *11*, 88. https://doi.org/10.3390/antiox110100.

Mantzarlis, K., et al. (2017). Role of oxidative stress and mitochondrial dysfunction in sepsis and potential therapies. *Oxidative Medicine and Cellular Longevity*, *2017*, 5985209. https://doi.org/10.1155/2017/5985209.

Maslove, D. M., et al. (2022). Redefining critical illness. *Nature Medicine*, *28*, 1141–1148. https://doi.org/10.1038/s41591-022-01843-x.

Merz, T. M., et al. (2017). Mitochondrial function of immune cells in septic shock: A prospective observational cohort study. *PLoS One*, *12*, 0178946. https://doi.org/10.1371/journal.pone.0178946.

Mohanty, T., et al. (2023). A pharmacoproteomic landscape of organotypic intervention responses in Gram-negative sepsis. *Nature Communications*, *14*, 3603. https://doi.org/10.1038/s41467-023-39269-9.

Nedeva, C., et al. (2019). Sepsis: Inflammation is a necessary evil. *Frontiers in Cell and Development Biology*, 7, 108. https://doi.org/10.3389/fcell.2019.00108.

Perelman, A., et al. (2012). JC-1: Alternative excitation wavelengths facilitate mitochondrial membrane potential cytometry. *Cell Death & Disease*, *3*, 430. https://doi.org/10.1038/cddis.2012.171.

Poll, V., et al. (2017). The immunopathology of sepsis and potential therapeutic targets. *Nature Reviews. Immunology*, *17*, 407–420. https://doi.org/10.1038/nri.2017.36.

Poll, V., et al. (2021). The immunology of sepsis. *Immunity*, *54*, 2450–2464. https://doi.org/10.1016/j.immuni.2021.10.012.

Potz, B. A., et al. (2016). Endothelial ROS and impaired myocardial oxygen consumption in sepsis-induced cardiac dysfunction. *Journal of Intensive and Critical Care*, *2*, 20. https://doi.org/10.21767/2471-8505.100020.

Ricquier, D., & Bouillaud, F. (2000). Mitochondrial uncoupling proteins: From mitochondria to the regulation of energy balance. *The Journal of Physiology*, *529*, 3–10. https://doi.org/10.1111/j.1469-7793.2000.00003.x.

Shenoy, S., et al. (2020). Coronavirus (Covid-19) sepsis: Revisiting mitochondrial dysfunction in pathogenesis, aging, inflammation, and mortality. *Inflammation Research*, *69*, 1077–1085. https://doi.org/10.1007/s00011-020-01389-z.

Singer, M. (2007). Mitochondrial function in sepsis: Acute phase versus multiple organ failure. *Critical Care Medicine*, *35*, S441–S448. https://doi.org/10.1097/01.CCM.0000278049.48333.78.

Singer, M. (2014). The role of mitochondrial dysfunction in sepsis-induced multi-organ failure. *Virulence*, *5*, 66–72. https://doi.org/10.4161/viru.26907.

Singer, M., et al. (2016). The third international consensus definition for sepsis and septic shock (sepsis-3). *Journal of the American Medical Association*, *315*, 801–810. https://doi.org/10.1001/jama.2016.0287.

Sivandzade, F., et al. (2019). Analysis of the mitochondrial membrane potential using the cationic JC-1 dye as a sensitive fluorescent probe. *Bio-Protocol*, *9*(1), e3128. https://doi.org/10.21769/BioProtoc.3128.

Sjövall, F., et al. (2013). Patients with sepsis exhibit increased mitochondrial respiratory capacity in peripheral blood immune cells. *Critical Care*, *17*, R152. https://doi.org/10.1186/cc12831.

Thompson, K., et al. (2018). Health-related outcomes of critically ill patients with and without sepsis. *Intensive Care Medicine*, *44*, 1249–1257. https://doi.org/10.1007/s00134-018-5274-x.

Víctor, V. M., et al. (2009). Oxidative stress and mitochondrial dysfunction in sepsis: A potential therapy with mitochondria-targeted antioxidants. *Infectious Disorders Drug Targets*, *9*, 376–389. https://doi.org/10.2174/187152609788922519.

Weiss, S. L., et al. (2015). Mitochondrial dysfunction in peripheral blood mononuclear cells in pediatric septic shock. *Pediatric Critical Care Medicine*, *54*, 285–293. https://doi.org/10.1097/SHK.0000000000001486.

Yang, Z., et al. (2014). Differential role for p120-catenin in regulation of TLR4 signaling in macrophages. *Journal of Immunology*, *193*, 1931–1941. https://doi.org/10.4049/jimmunol.1302863.

Yao, X., et al. (2015). Mitochondrial ROS induces cardiac inflammation via a pathway through mtDNA damage in a pneumonia-related sepsis model. *PLoS One*, *10*, e0139416. https://doi.org/10.1371/journal.pone.0139416.

Zhang, H., et al. (2018). Potential therapy strategy: Targeting mitochondrial dysfunction in sepsis. *Military Medical Research*, *51*, 41. https://doi.org/10.1186/s40779-018-0187-0.

Zhu, C. L., et al. (2021). Mechanism of mitophagy and its role in sepsis-induced organ dysfunction: A review. *Frontiers in Cell and Development*, *9*. https://doi.org/10.3389/fcell.2021.664896.

Zorova, L. D., et al. (2018). Mitochondrial membrane potential. *Analytical Biochemistry*, *552*, 50–59. https://doi.org/10.1016/j.ab.2017.07.009.

CHAPTER FOUR

Quantification of mitochondrial reactive oxygen species in macrophages during sepsis

Kanmani Suganya[a], Paulraj Kanmani[a], and Guochang Hu[a,b,*]

[a]Department of Anesthesiology, University of Illinois College of Medicine, Chicago, IL, United States

[b]Department of Pharmacology & Regenerative Medicine, University of Illinois College of Medicine, Chicago, IL, United States

*Corresponding author: e-mail address: gchu@uic.edu

Contents

Abstract

Sepsis is the leading causes of death globally, arising from an imbalanced host response to severe infection. It leads to multi-organ failure and poor outcomes in septic patients due to compromised glucose and lipid oxidation, reduced oxygen consumption, elevated levels of circulating substrates, and impaired mitochondrial function. Mitochondria, essential cellular organelles, play a vital role in regulating various cellular

Methods in Cell Biology, Volume 194
ISSN 0091-679X
https://doi.org/10.1016/bs.mcb.2024.01.005

activities and the host immune response to infection. Pathogens, particularly bacteria, often disrupt mitochondrial functions to dysregulate host immunity. Additionally, the mitochondrial function is closely associated with most host immune responses, making mitochondria crucial in maintaining host homeostasis during infection. The intrinsic inflammatory response triggered by pathogens in sepsis impairs mitochondrial function, resulting in excessive production of mitochondrial reactive oxygen species (ROS) and subsequently damage to multiple organs. Here, we present a simple protocol for assessing mitochondrial ROS levels in bone marrow-derived macrophages (BMDMs) isolated from mice. We observed a higher level of ROS generation in lipopolysaccharide (LPS)-treated BMDMs, indicating the effectiveness and efficiency of our designed protocol for assessing mitochondrial ROS generation *in vitro*.

1. Introduction

Sepsis, an imbalanced and life-threatening systemic inflammation caused by infection, is the leading cause of death, with a rising number of reported deaths among patients admitted to intensive care units (Singer et al., 2016; Vincent et al., 2006). Multiorgan failure accounts for 80% of deaths in patients with severe sepsis and septic shock (Fleischmann et al., 2016; Marshall et al., 2005). Various factors, including mitochondrial dysfunction, energy depletion, impaired vascular permeability, and cardiac malfunction, contribute to the pathogenesis of sepsis (Mantzarlis et al., 2017; Nagar et al., 2018). Of these factors, mitochondrial dysfunction plays a crucial role in the development of vital organ failure in sepsis (Brealey et al., 2002; Chen et al., 2017; Fialkow et al., 2007).

Mitochondria, the predominant cellular organelles with a high capacity for producing reactive oxygen species (ROS), are central to cellular processes (Joseph, 2017; Tiku, 2020). ROS comprise a group of molecules, which include superoxide radicals, hydroxyl radicals, peroxide radicals, and singlet oxygen. Within mitochondria, the main ROS produced is superoxide. Superoxide is formed through single electron transfer to O_2 by several enzymes of the electron transport chain including Complexes I and III, as well as glycerol-3-phosphate dehydrogenase, and serves as the common precursor for all ROS produced by cells (Hayyan et al., 2016). During oxidative phosphorylation, the respiratory chain located in the inner membrane of mitochondria generates ROS (Reichart et al., 2019; Scialo et al., 2017). Physiological levels of ROS play essential roles in maintaining internal homeostasis and regulating various functions such as host defenses, cellular signaling, proliferation, redox regulation, signal transduction, and mitochondrial biogenesis. However, pathophysiological levels of ROS contribute to disease progression, including sepsis (Droge, 2002; Lee & Wei, 2005;

Nagar et al., 2018; Preau et al., 2021; Ray et al., 2012). Excessive production of ROS can result in mitochondrial damage, including impairment of mitochondrial structure and biogenesis due to oxidative modification of macromolecules, mutations in mitochondrial DNA, damage to the mitochondrial respiratory chain and mitochondrial membrane permeability, as well as disruption of Ca^{2+} homeostasis (Shokolenko et al., 2009; Tsutsui et al., 2008).

The mitochondrion serves as a significant cellular source of ROS in infected immune cells, including macrophages. Under normal, pathogen-free conditions, mitochondria generate minimal levels of ROS. However, during sepsis, the recognition of pathogens by pattern recognition receptors such as Toll like receptors, can induce excessive production of mitochondrial ROS in immune cells (Herb & Schramm, 2021). Proinflammatory cytokines, including interferon γ (IFNγ) and tumor necrosis factor (TNF)α, have been shown to stimulate mitochondrial ROS production in infected immune cells (Roca et al., 2019; Roca & Ramakrishnan, 2013; Sonoda et al., 2007). Mitochondrial ROS play a significant role in the interplay between mitochondria and bacteria. Mitochondria limit the survival and dissemination of intracellular bacteria in the host through the production of ROS. Bacteria frequently target and disrupt mitochondrial functions, triggering the generation of ROS (Escoll et al., 2016; Marchi et al., 2022). Research conducted both by our laboratory (Kanmani et al., 2023) and by others (Cai et al., 2023; Maher et al., 2014) has demonstrated that lipopolysaccharide (LPS), the major cell wall component of Gram-negative bacteria, amplifies mitochondrial ROS production in bone marrow-derived macrophages (BMDMs). Mitochondrial ROS in immune cells play a vital dual role in modulating inflammatory responses during sepsis. One beneficial function of mitochondrial ROS is their antimicrobial activity in host defense (Shekhova, 2020). Additionally, mitochondrial ROS influence the production rate of cytokines during infection, thereby playing a crucial role in regulating inflammation *in vivo* (Garaude et al., 2016; Mills et al., 2016). However, despite the essential role of mitochondrial ROS in immune cell antimicrobial responses, overproduction of mitochondrial ROS may exhibit detrimental properties and contribute to host damage during infection. Specifically, mitochondrial ROS have been implicated in organ failure in various sepsis models (Lowes et al., 2008; Ramsey & Wu, 2014). Moreover, mitochondrial ROS could potentially contribute to heightened immune responses during viral infections, including those caused by the influenza A virus (To et al., 2020). In activated immune cells, the production of mitochondrial ROS leads to membrane permeability transition, resulting in the release of mitochondrial DNA into the cytosol. This, in turn, elevates the concentration of IL-1β and exaggerates inflammatory response and tissue injury (Kanmani et al., 2023; Nakahira et al., 2011). In summary, mitochondrial ROS produced in immune cells play a pivotal role in regulating immune homeostasis in the context of sepsis. Therefore, it is of utmost importance to precisely detect and quantify mitochondrial ROS generation both *in vitro* and *in vivo* during infection.

There is currently an abundance of commercially available ROS probes that enable ROS measurements in *ex vivo* cells. The commonly employed ROS probes exhibit either chemiluminescence emission through a ROS-consuming reaction or fluorescence emission following their reaction with ROS. One such probe is MitoSOX Red, which is a modified dihydroethidium (DHE) analog attached to a triphenylphosphonium (TPP) group. This attachment facilitates the accumulation of the probe specifically in the mitochondrial matrix (Mukhopadhyay et al., 2007; Robinson et al., 2006). MitoSOX Red is a fluorescent probe that can permeate live cells and rapidly and selectively target mitochondria. Once inside the mitochondria, MitoSOX Red undergoes oxidation specifically by superoxide, but not by other ROS or reactive nitrogen species (RNS) generating systems. This oxidation results in the probe binding to intramitochondrial nucleic acids, leading to the generation of intense red fluorescence. Thus, MitoSOX Red serves as a specific fluorescent indicator for detecting mitochondrial superoxide. It specifically localizes to the mitochondrial matrix, making it a reliable indicator for superoxide detection (Herb et al., 2019; Wang & Zou, 2018). Rotenone is a commonly used positive control to induce mitochondrial ROS generation and, consequently, MitoSOX Red fluorescence (Herb et al., 2019; Li et al., 2003; Wu et al., 2013; Zhou et al., 2011). The assay, which can be completed within 1 h, is easily detectable through fluorescence microscopy. It offers a sensitive and one-step fluorometric method to detect mitochondrial superoxide radicals in cultured cells. In this study, we present quantitative techniques for measuring the rate of mitochondria ROS production in BMDMs using the specific indicator, MitoSOX Red.

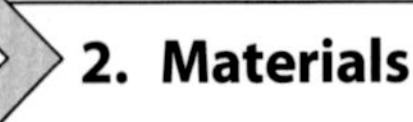

2. Materials

2.1 Common disposables

- 12 well cell culture plate collagen coated (Falcon, #353043) (see *Notes 1 and 2*)

- 6 well tissue culture plate collagen coated (Denville, #1020A) (see *Notes 1 and 2*)
- 60 mm tissue culture dish (Denville, #T1116) (see *Notes 1 and 2*)
- 100 mm tissue culture dish (Denville, T1110) (see *Notes 1 and 2*)
- Disposable centrifuge tube (15 mL); sterile (Corning Science, #352096) (see *Notes 1 and 2*)
- Disposable centrifuge tube (50 mL); sterile (Fisher Scientific, #05-539-7) (see *Notes 1 and 2*)
- Cell strainer (70 μm pore size), sterile (Falcon, #351029) (see *Notes 1 and 3*)
- Syringe filter (0.2 μm, pore size, #4652, PALL Life Sciences) (see *Notes 1 and 3*)
- 10 mL Syringe (BD, #303134) (see *Note 4*)
- 25-gauge needle, sterile (BD, #305122) (see *Note 4*)

2.2 Surgical instruments

- Dissection scissors (Fisherbrand #08-940) (See *Note 1*)
- Curved medium forceps (Fisherbrand, #16-100-110) (See *Note 1*)
- General purpose tweezers (Fisherbrand #17-467-235) (See *Note 1*)

2.3 Cells and reagents

- L929 cells; NCTC clone 929 Areolar fibroblasts; Mouse (CCL-1, ATCC, Rockville, MD) (see *Note 5*)
- Dulbecco's modified Eagle's medium (DMEM; #11965-092, Gibco) (see *Notes 1 and 6*)
- Roswell Park Memorial Institute (RPMI) 1640 medium (#11875-093, Gibco; USA) (see *Notes 1 and 6*)
- Fetal bovine serum (FBS; #26140079, Gibco) (see *Notes 1 and 6*)
- Penicillin and 100 μg/mL streptomycin (10,000 U/mL, #15140122, Gibco) (see *Notes 1 and 6*)
- On-TargetPlus Mouse Ctnnd1 siRNA SmartPool (10 nmol; #L-062106-01-0010; Dharmacon)
- Phosphate buffered saline (PBS; Corning, #21-040-CV) (see *Note 1*)
- Lipopolysaccharide (LPS; InVivogen, #NC0202558) (see *Notes 1 and 7*)
- 70% ethanol (70 mL of ethanol mixed with 30 mL of water) (see *Notes 1 and 8*)
- MitoSOX Red (#M36008; InVitrogen, USA) (see *Notes 1 and 9*)
- Gelatin (#G1393, Sigma, USA) (see *Note 1*)
- Formaldehyde (#1004960700, Sigma) (see *Note 1*)

- ATP (#A6419, Sigma) (see *Note 1*)
- Non-enzymatic cell dissociation solution (#27420004; Corning, USA) (see *Note 1*)
- ProLong™ Gold Antifade Mountant (#P36934; InVitrogen, USA)

2.4 Equipment

- Laboratory biosafety cabinet (SterilGard Hood, USA) (see *Note 1*)
- Humidified cell culture incubator sets at 37 °C with 5% CO_2 (Fisher Scientific, Model: 3530) (see *Note 1*)
- Inverted Fluorescence microscope (Olympus Life Sciences, USA) (see *Note 1*)
- Standard bench top centrifuge (Eppendorf, Centrifuge 5415R) (see *Note 1*)
- Eisco thick-walled mortar and pestle (#S39830, Fisher Scientific, USA)

2.5 Software

- Prism software (see *Notes 1 and 10*)
- ImageJ software (see *Notes 1 and 10*)

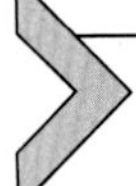

3. Methods

3.1 Maintenance of mice

1. C57BL/6J mice were obtained from Jackson Laboratory and housed in a specific pathogen-free barrier facility. The animal protocols were approved by the Institutional Animal Care and Use Committee of the University of Illinois at Chicago (see *Note 11*)
2. All mice had free access to water and food. Experiments were conducted using age-matched (12–16 weeks old) male and female mice in a 1:1 ratio (see *Note 12*)
3. Mice were anesthetized using either 1–3% isoflurane *via* inhalation or ketamine/xylazine (100 mg/kg/12.5 mg/kg, i.p.) injection (see *Note 13*)

3.2 Culture of L929 cells

1. Culture L929 cells at a density of 5×10^4 cells/mL in DMEM medium supplemented with 10% fetal bovine serum, 100 U/mL penicillin, and 100 μg/mL streptomycin. Incubate the cells in a humidified incubator at 37 °C with 5% CO_2 (see *Notes 5 and 14*)

2. On day 3, carefully collect the cultured medium without disturbing the confluent cells attached to the bottom of the culture dish. Filter the medium using a syringe filter with a pore size of 0.2 μm to remove any dead or live cells present in the supernatant
3. Transfer the filtered culture medium into a 50 mL Falcon tube and store it at −80 °C for the preparation of macrophage differentiation medium (see *Notes 14 and 15*)

3.3 Isolation of bone marrow-derived macrophages from mice

1. Euthanize C57BL/6 mice by intraperitoneal injection of 3 mg/kg xylazine and 75 mg/kg ketamine, followed by cervical dislocation (see *Note 16*)
2. Sterilize the skin with 70% ethanol, make an incision in the abdomen, and sever the portal vein to ensure complete removal of blood from the mouse's body (see *Note 17*)
3. Use sterile surgical scissors and forceps to remove the skin and expose the lower limb of the legs
4. Remove the surrounding flesh and tissues around the bones to expose the hip joint (see *Note 18*)
5. Cut at the knee and hip joints and transfer the bones to a dish containing 5–8 mL of sterile 1× ice-cold PBS (see *Note 18*)
6. Place the bones in a sterile cell culture hood and carefully remove any additional tissues that may be adhered to the femur and tibia (see *Note 19*)
7. After removing the tissues, rinse the bones with sterile 1× ice-cold PBS and sterilize them by transferring them to a sterile dish containing 70% ethanol for 2 min (see *Note 19*)
8. Transfer the bones to a new dish containing RPMI complete medium and separate the tibia and femur by cutting the joint between them
9. Place the two pieces of bones (tibia and femur) in a mortar containing 5 mL of RPMI medium (see *Note 20*)
10. Gently crush both the tibia and femur using a sterile pestle, collecting the supernatant in a 50 mL Falcon tube. Repeat this step at least 2–3 times (see *Note 21*)
11. Filter the supernatant using a 70-μm pore size cell strainer and centrifuge the filtrate at 300 × *g* for 5 min at room temperature (see *Notes 4 and 22*)

12. Discard the supernatant and add 1 mL of red cell lysis buffer to the pellet to lyse any remaining red blood cells (RBCs). Repeat this step if RBCs are still present in the pellet.
13. Pipette up and down for 30–40 s, add 15 mL of RPMI medium, and centrifuge at 300 × *g* for 5 min at RT (see *Notes 1 and 23*)

3.4 Culture of bone marrow-derived macrophages

1. Aspirate the supernatant and wash the cells with PBS. Dissociate the pellet in 20 mL of warm macrophage differentiation culture medium prepared by mixing RPMI complete medium and L929 cultured supernatants at a 7:3 ratio (see *Note 24*)
2. Count the cells and plate 10 mL of cells (approximately 1×10^7 cells/mice) into a sterile 100 mm tissue culture dish (see *Note 25*)
3. Incubate the cells in the culture dish in a humidified incubator at 37 °C with 5% CO_2 for 2 h (see *Note 26*)
4. Transfer the medium with non-adherent cells to a new dish and incubate the cells in a humidified incubator at 37 °C with 5% CO_2. Discard the plate containing stromal cells (see *Note 27*)
5. Allow the cells to grow for 3 days and add 5 mL of fresh differentiation medium to the culture dish. Incubate the cells at 37 °C with 5% CO_2 for an additional 2 days (see *Note 27*)
6. On day 5, add 5 mL of fresh differentiation medium to the culture dish and continue the incubation at 37 °C with 5% CO_2 for another 1 day (see *Note 27*)
7. On day 6, aspirate the culture medium and add fresh 10 mL of differentiation medium to the culture dish. Continue the incubation at 37 °C with 5% CO_2 for another 24 h
8. The following day, mature macrophages will be firmly attached to the bottom of the culture dish, ready for analysis of mitochondrial ROS production *in vitro* (see *Note 28*)

3.5 Small interfering RNA transfection

1. Culture the mature BMDMs (1×10^6) in a sterile 6-well tissue culture plate for 24 h (see *Note 29*)
2. Prepare the transfection reagent by mixing scrambled (Sc) small interfering RNA (siRNA) or p120 catenin (p120) siRNA (20 ng) at a

concentration of 50 nM, RNA iMax (8 μL), and serum-free medium (280 μL) (see *Note 30*)

3. Add 300 μL of the transfection reagent to each well containing serum-free RPMI medium and incubate at 37 °C, 5% CO_2 for 24 h
4. Remove the old medium and replace it with fresh RPMI complete medium in each well for an additional 24 h (see *Note 31*)
5. Verify the effectiveness of p120 knockdown through Western blot analysis

3.6 Preparation of imaging cell culture slide

1. Take the desired size and number of glass cover slips and place them in a 100 mL glass beaker with a layer of cotton at the bottom. Wrap the beaker with aluminum foil and sterilize it at 120 °C for 1 h
2. Place the sterile glass cover slips individually in a 12-well plate. Treat each cover slip with gelatin for 2 min, then wash it by adding 1.0 mL of fresh RPMI medium (see *Note 32*)
3. After knockdown, wash the cells with sterile PBS and collect them by adding 500 μL of non-enzymatic solution to the well (see *Note 32*)
4. Transfer the cells to a 15 mL tube containing 10 mL of fresh RPMI medium and centrifuge at 1500 rpm for 5 min (see *Note 32*)
5. Remove the supernatant and wash the cells three times with 1 mL of PBS (see *Note 32*)
6. Dissolve the cell pellets in 1.0 mL of fresh RPMI medium and count the cells (see *Note 32*)
7. Add the desired number of both control and p120-knockdown BMDMs to the 12-well plate containing the gelatin-treated glass cover slips. Place the plate in an incubator set at 37 °C, 5% CO_2 (see *Note 32*)
8. Allow the cells to attach to the surface of the cover slips overnight

3.7 Determination of mitochondrial ROS production

1. Prepare LPS at a concentration of 20 ng/mL in RPMI medium in a 15 mL Falcon tube. Prepare the required volume of medium based on the number of wells. Add 1.0 mL of LPS-containing medium to each well and incubate at 37 °C, 5% CO_2 for 4 h. After that, treat with ATP (5 mM) for 30 min (see *Note 33*)

2. Prepare 5 mM MitoSOX Red in a 15 mL Falcon tube by adding the appropriate volume of RPMI medium. Cover the tube with aluminum foil until ready to use (see *Note 11*)
3. Remove the medium from the wells and add 1.0 mL of MitoSOX Red medium to each well. Continue incubating at 37 °C, 5% CO_2 for 30 min (see *Note 11*)
4. Remove the medium and wash the cells three times with warm serum-free RPMI medium, followed by PBS
5. Fix the cells on the glass cover slips by adding 1.0 mL of 4% formaldehyde for 15 min. This step is preferably performed in the dark
6. Remove the formaldehyde and wash the slides three times with PBS. Then, add 1.0 mL of distilled H_2O to each well. Keep the plates in the dark until the slides are mounted (see *Note 34*)
7. Take the glass cover slips out of the wells, gently remove excess water by dabbing with a tissue, and mount them on clean glass slides with a drop of mounting medium (Prolong Gold Antifade Mountant). Allow the slides to dry and store them in the dark until examination under a microscope (see *Note 34*)
8. The production of mitochondrial ROS can be visualized by fluorescence emission, which can be captured using an inverted fluorescence microscope at 60× magnification. The intensity of fluorescence can be quantified using ImageJ software (NIH). It is advisable to use a confocal microscope to capture and measure mitochondrial ROS production in the cells (see *Notes 35 and 36*)

3.8 Data analysis

1. BMDMs were treated with LPS and subsequently incubated with MitoSOX Red. After fixation, the cells were analyzed under the microscope, and the results are presented in Fig. 1A and B
2. ImageJ software was utilized for quantifying the fluorescence intensity, while Prism software was employed for data analysis, as well as generating bar and scatter plots (see *Note 12*)
3. Statistical analysis involved employing one-way ANOVA with Bonferroni *post hoc* test to assess the presence of statistically significant differences between the experimental groups

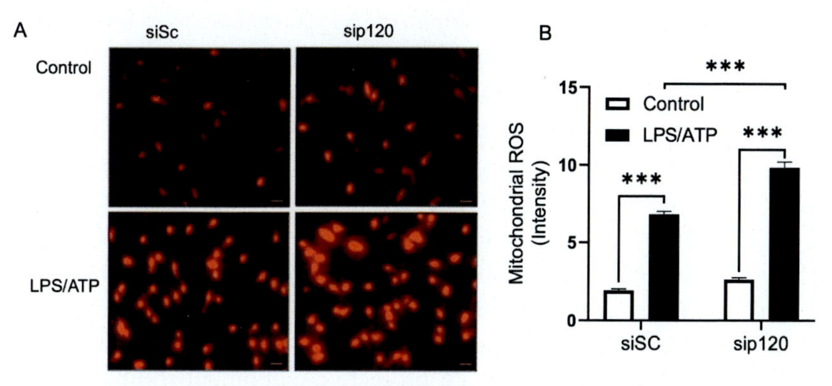

Fig. 1 Measurement of mitochondrial ROS production in macrophages in response to LPS stimulation *in vitro*. BMDMs were isolated from C57BL/6J mice and cultured according to the methods described in Sections 3.2–3.4. Subsequently, mature BMDMs were transfected with scrambled (siSc) and p120 siRNA (sip120), as outlined in Section 3.5. After 48 h of transfection, the transfected cells were incubated with LPS (20 ng/mL) for 4 h and then stimulated with ATP (5 mM) for 30 min. This was followed by incubation with MitoSOX Red (5 mM) for an additional 30 min. (A) Representative images of mitochondrial ROS in BMDMs are shown, with a scale bar of 10 μm. The images for both the control and treatment groups were collected simultaneously and under identical conditions. (B) The MitoSOX Red fluorescence intensity was quantified ($n=6$), and the fluorescence signal was analyzed using Scion Image software. The data presented represent means ± SD ($n=3$), with n indicating the number of experimental replicates. $^{***}P<0.001$ (one-way ANOVA with Bonferroni *post hoc* test) (see ref. Kanmani et al., 2023).

4. Notes

1. The catalog numbers of companies and their products are provided as a reference for readers, and it is not suggested to exclusively purchase or use products from those specific companies for your experiments. Similar products or reagents are available from several other companies.
2. Various culture dishes or plates with different coating systems can be used, and they are readily available from multiple companies.
3. Different companies offer cell strainers and syringe filters, which should be sterile.
4. Different sizes of syringes and needles can be utilized for this experiment.

5. L929 cells (CLL-1) are available from ATCC and are used for the production of colony-stimulating factors (CSF) that aid in the differentiation of progenitor cells into mature BMDMs. Several commercially available CSFs are also accessible.
6. Growth medium, FBS, and PBS are non-hazardous. However, it is recommended to use protective gloves to prevent contamination. It is advisable to use antibiotics at the same concentration, and caution should be exercised when handling these products. The use of antifungal antibiotics is also recommended.
7. Lipopolysaccharide (LPS), an inflammatory agent, can be harmful to health. Therefore, appropriate protective equipment should be used when working with LPS.
8. Ethanol is a highly flammable liquid and may cause serious eye irritation. Caution should be exercised when handling ethanol.
9. MitoSOX Red can be purchased from various manufacturers. It is typically supplied in 10 × 50 µg vials, which should be dissolved with 13 µL of Dimethylsulfoxide (DMSO) to prepare a stock solution of 50 mM. The prepared vials should be stored at −20 °C. It is recommended to wear appropriate protective gloves when working with MitoSOX Red, and the prepared solution should be kept in the dark. The MitoSOX Red reagent, being a derivative of ethidium bromide, should be handled with the necessary caution and care.
10. Software versions are indicated as a reference, but equivalent software can be obtained from various providers.
11. C57BL/6J is a wild-type mouse strain widely used by researchers for their experiments. It must be maintained in a specific pathogen-free facility. Other mouse strains can be used for the isolation of BMDMs.
12. The age of mice can vary, but it is advisable to use mice aged 10–12 weeks. Mice should be fed *ad libitum* at the time of sacrificing and extracting BMDMs.
13. The concentration and volume of anesthetic drugs can vary; however, the mentioned concentrations work well and are sufficient for achieving mouse anesthesia.
14. The growth medium should be brought to room temperature, and all cell culture steps should be performed under a Class II Biological Safety Cabinet to prevent bacterial and fungal contamination.
15. L929 cells can be grown in other types of tissue culture plates, and culturing for 3 days is recommended, although longer durations are also acceptable. The cell culture supernatant should be filtered to eliminate debris and dead cells and stored at −70 °C.

16. Strict adherence to approved animal ethics protocols is essential when euthanizing mice.
17. Different methods can be employed to remove blood from mice before extracting the legs for BMDM isolation.
18. Tissues must be completely removed to expose the bones for viewing.
19. Once the bones are removed from the mice, they should be immediately taken into a biosafety cabinet hood to remove any remaining tissues surrounding them.
20. The mortar should be in a sterile condition and the process should be performed within a biosafety cabinet hood.
21. The bones should be thoroughly crushed using a pestle, and the bone marrow should be dissociated using a 25G needle. Researchers may choose to use a syringe fitted with an appropriate needle to extract bone marrow from the bones instead of crushing the bones in a mortar.
22. The cell suspension should be filtered through a cell strainer to remove tissues and bones.
23. Red blood cells should be lysed, and the lysis step should be repeated until complete lysis is achieved. Complete removal of red blood cells will increase the yield of BMDMs and enhance precursor populations.
24. The cells should be washed with PBS, and the cell pellets should be mixed with differentiation medium before transferring them to plates. Reagents used in this experiment can be replaced with alternative products, especially the medium. For instance, the cells can be cultured in DMEM medium as a substitute for RPMI medium.
25. Cells can be plated in any tissue culture dish, and any remaining cells can be stored in liquid nitrogen for future use.
26. The cells can be incubated for more than 2 h if necessary.
27. Stromal cells should be removed, and only non-adherent BMDMs should be transferred to fresh cell culture plates.
28. Fresh medium should be added to the dish every 2 days for a total of 6 days. Mature cells should be used for the experiment and mitochondrial ROS production.
29. The cells should be seeded in a 6- or 12-well plate for the experiment. Enzyme-free cell dissociation buffer should be used to detach cells from the dish.
30. The cells should be grown in serum-free medium during siRNA transfection. Care should be taken to avoid air bubbles while pipetting the transfection reagents.

31. Once the old medium is removed, fresh medium should be added immediately, and the cells should continue to grow for another 12 or 24h.
32. Sterilized glass cover slides should be used to prevent contamination. The cover slides should be treated with gelatin to ensure firm cell attachment.
33. LPS can be purchased from different companies. The concentration and duration of LPS treatment used in this study are optimized for the specific cells. However, the concentration and duration may vary depending on the cell type and mice.
34. Avoid exposing the cells to light while preparing them for the analysis of mitochondrial ROS production. Any water on the slides should be removed from the slips before mounting. If nucleus staining is required, use a mounting medium with DAPI.
35. The cells can be washed with PBS instead of medium.
36. It is important to quantify ROS levels as soon as the cells are treated with LPS since the lifespan of most ROS is short and their steady-state levels are low.

5. Concluding remarks

The crucial and multifaceted roles of mitochondrial ROS in macrophage-mediated immunity and inflammatory responses during sepsis are widely recognized and have been validated by numerous studies (Droge, 2002; Kanmani et al., 2023; Lee & Wei, 2005; Nagar et al., 2018; Preau et al., 2021; Ray et al., 2012). However, it is important to note that experimental setups that employ inappropriate mitochondrial ROS probes can often result in misinterpretation of the measurements in macrophages. MitoSOX Red is one of the widely utilized probes for detecting mitochondrial ROS, and MitoSOX is commonly employed due to its ability to accumulate within mitochondria as a result of its positive charge. MitoSOX-based assays have been extensively applied for measuring mitochondrial ROS content in cells across diverse conditions. The protocol described in this study presents a straightforward and highly sensitive method utilizing MitoSOX Red for quantifying mitochondrial ROS content in BMDMs using a fluorescence microscope. BMDMs, derived from *in vitro* differentiation of bone marrow cells with the culture medium conditioned by the L929 murine fibroblast cell line contains macrophage colony-stimulating factor (M-CSF), serve as primary macrophages. These cells can be obtained in ample quantities, possess the functional characteristics of mature macrophages, and can be stored *via* freezing. While this protocol specifically focuses on quantifying mitochondrial ROS content in BMDMs *in vitro*, it can be adapted for use with various types of mammalian cells.

Acknowledgments

The work was supported by an R01 grant (#HL152696, PI: G.H.) and an R01 grant (#HL104092, PI: G.H.) from the National Heart, Lung, and Blood Institute of the National Institutes of Health, and an R21 grant (#AI152249, PI: G.H.) from the National Institute of Allergy and Infectious Diseases of the National Institutes of Health.

Competing interests

The authors have no conflicts of interest to declare.

References

Brealey, D., et al. (2002). Association between mitochondrial dysfunction and severity and outcome of septic shock. *Lancet, 360*, 219–223. https://doi.org/10.1016/S0140-6736(02)09459-X.

Cai, S., et al. (2023). Mitochondrial dysfunction in macrophages promotes inflammation and suppresses repair after myocardial infarction. *Journal of Clinical Investigation, 133*, e159498. https://doi.org/10.1172/JCI159498.

Chen, G., et al. (2017). Reactive oxygen species-responsive polymeric nanoparticles for alleviating sepsis-induced acute liver injury in mice. *Biomaterials, 144*, 30–41. https://doi.org/10.1016/j.biomaterials.2017.08.008.

Droge, W. (2002). Free radicals in the physiological control of cell function. *Physiological Reviews, 82*, 47–95. https://doi.org/10.1152/physrev.00018.2001.

Escoll, P., et al. (2016). MAMs are attractive targets for bacterial repurposing of the host cell. *BioEssays, 39*, 1600171. https://doi.org/10.1002/bies.201600171.

Fialkow, L., et al. (2007). Reactive oxygen and nitrogen species as signaling molecules regulating neutrophil function. *Free Radical Biology and Medicine, 42*, 153–164. https://doi.org/10.1016/j.freeradbiomed.2006.09.030.

Fleischmann, C., et al. (2016). Assessment of global incidence and mortality of hospital-treated sepsis. Current estimates and limitations. *American Journal of Respiratory and Critical Care Medicine, 193*, 259–272. https://doi.org/10.1164/rccm.201504-0781oc.

Garaude, J., et al. (2016). Mitochondrial respiratory-chain adaptations in macrophages contribute to antibacterial host defense. *Nature Immunology, 17*, 1037–1045. https://doi.org/10.1038/ni.3509.

Hayyan, M., et al. (2016). Superoxide ion: Generation and chemical implications. *Chemical Reviews, 116*, 3029–3085. https://doi.org/10.1021/acs.chemrev.5b00407.

Herb, M., & Schramm, M. (2021). Functions of ROS in macrophages and antimicrobial immunity. *Antioxidants (Basel), 10*, 313. https://doi.org/10.3390/antiox10020313.

Herb, M., et al. (2019). Mitochondrial reactive oxygen species enable proinflammatory signaling through disulfide linkage of NEMO. *Science Signaling, 12*, eaar5926. https://doi.org/10.1126/scisignal.aar5926.

Joseph, L. C., et al. (2017). Inhibition of NADPH oxidase 2 (NOX2) prevents sepsis-induced cardiomyopathy by improving calcium handling and mitochondrial function. *Journal of Clinical Investigation, 2*, e94248. https://doi.org/10.1172/jci.insight.94248.

Kanmani, P., et al. (2023). p120-Catenin suppresses NLRP3 inflammasome activation in macrophages. *American Journal of Physiology. Lung Cellular and Molecular Physiology, 324*, L596–L608. https://doi.org/10.1152/ajplung.00328.2022.

Lee, H. C., & Wei, Y. H. (2005). Mitochondrial biogenesis and mitochondrial DNA maintenance of mammalian cells under oxidative stress. *International Journal of Biochemistry & Cell Biology, 37*, 822–834. https://doi.org/10.1016/j.biocel.2004.09.010.

Li, N., et al. (2003). Mitochondrial complex I inhibitor rotenone induces apoptosis through enhancing mitochondrial reactive oxygen species production. *Journal of Biological Chemistry, 278*, 8516–8525. https://doi.org/10.1074/jbc.M210432200.

Lowes, D. A., et al. (2008). The mitochondria-targeted antioxidant MitoQ protects against organ damage in a lipopolysaccharide-peptidoglycan model of sepsis. *Free Radical Biology and Medicine, 45*, 1559–1565. https://doi.org/10.1016/j.freeradbiomed.2008.09.003.

Maher, K., et al. (2014). A role for stefin B (cystatin B) in inflammation and endotoxemia. *Journal of Biological Chemistry, 289*, 31736–31750. https://doi.org/10.1074/jbc.M114.609396.

Mantzarlis, K., et al. (2017). Role of oxidative stress and mitochondrial dysfunction in sepsis and potential therapies. *Oxidative Medicine and Cellular Longevity*, 5985209. https://doi.org/10.1155/2017/5985209.

Marchi, S., et al. (2022). Control of host mitochondria by bacterial pathogens. *Trends in Microbiology, 30*, 452–465. https://doi.org/10.1016/j.tim.2021.09.010.

Marshall, J. C., et al. (2005). Outcome measures for clinical research in sepsis: A report of the 2nd Cambridge Colloquium of the International Sepsis Forum. *Critical Care Medicine, 33*, 1708–1716. https://doi.org/10.1097/01.ccm.0000174478.70 338.03.

Mills, E. L., et al. (2016). Succinate dehydrogenase supports metabolic repurposing of mitochondria to drive inflammatory macrophages. *Cell, 167*. https://doi.org/10.1016/j.cell.2016.08.064. 457–470.e13.

Mukhopadhyay, P. (2007). Simultaneous detection of apoptosis and mitochondrial superoxide production in live cells by flow cytometry and confocal microscopy. *Nature Protocols, 2*, 2295–2301. https://doi.org/10.1038/nprot.2007.327.

Nagar, H., et al. (2018). Role of Mitochondrial oxidative stress in sepsis. *Acute Critical Care, 33*, 65–72. https://doi.org/10.4266/acc.2018.00157.

Nakahira, K., et al. (2011). Autophagy proteins regulate innate immune responses by inhibiting the release of mitochondrial DNA mediated by the NALP3 inflammasome. *Nature Immunology, 12*, 222–230. https://doi.org/10.1038/ni.1980.

Preau, S., et al. (2021). Energetic dysfunction in sepsis: A narrative review. *Annals of Intensive Care, 11*, 104. https://doi.org/10.1186/s13613-021-00893-7.

Ramsey, H., & Wu, M. X. (2014). Mitochondrial anti-oxidant protects IEX-1 deficient mice from organ damage during endotoxemia. *International Immunopharmacology, 23*, 658–663. https://doi.org/10.1016/j.intimp.2014.10.019.

Ray, P. D., et al. (2012). Reactive oxygen species (ROS) homeostasis and redox regulation in cellular signaling. *Cellular Signalling, 24*, 981–990. https://doi.org/10.1016/j.cellsig.2012.01.008.

Reichart, G., et al. (2019). Mitochondrial complex IV mutation increases reactive oxygen species production and reduces lifespan in aged mice. *Acta Physiology, 225*, e13214. https://doi.org/10.1111/apha.13214.

Robinson, K. M., et al. (2006). Selective fluorescent imaging of superoxide in vivo using ethidium-based probes. *Proceedings of the National Academy of Sciences of the United States of America, 103*, 15038–15043. https://doi.org/10.1073/pnas.0601945103.

Roca, F. J., & Ramakrishnan, L. (2013). TNF dually mediates resistance and susceptibility to mycobacteria via mitochondrial reactive oxygen species. *Cell, 153*, 521–534. https://doi.org/10.1016/j.cell.2013.03.022.

Roca, F. J., et al. (2019). TNF induces pathogenic programmed macrophage necrosis in tuberculosis through a mitochondrial-lysosomal-endoplasmic reticulum circuit. *Cell*, *178*. https://doi.org/10.1016/j.cell.2019.08.004. 1344–1361.e11.

Scialo, F., et al. (2017). Role of mitochondrial reverse electron transport in ROS signaling: Potential roles in health and disease. *Frontiers in Physiology*, *8*, 428. https://doi.org/10.3389/fphys.2017.00428.

Shekhova, E. (2020). Mitochondrial reactive oxygen species as major effectors of antimicrobial immunity. *PLoS Pathogens*, *16*, e1008470. https://doi.org/10.1371/journal.ppat.1008470.

Shokolenko, I., et al. (2009). Oxidative stress induces degradation of mitochondrial DNA. *Nucleic Acids Research*, *37*, 2539–2548. https://doi.org/10.1093/nar/gkp100.

Singer, M., et al. (2016). The third international consensus definitions for sepsis and septic shock (Sepsis-3). *Journal of the American Medical Association*, *315*, 801–810. https://doi.org/10.1001/jama.2016.0287.

Sonoda, J., et al. (2007). Nuclear receptor ERRα and coactivator PGC-1β are effectors of IFN-γ-induced host defense. *Genes & Development*, *21*, 1909–1920. https://doi.org/10.1101/gad.1553007.

Tiku, V., et al. (2020). Mitochondrial functions in infection and immunity. *Trends in Cell Biology*, *30*, 263–275. https://doi.org/10.1016/j.tcb.2020.01.006.

To, E. E., et al. (2020). Mitochondrial reactive oxygen species contribute to pathological inflammation during influenza A virus infection in mice. *Antioxidants & Redox Signaling*, *32*, 929–942. https://doi.org/10.1089/ars.2019.7727.

Tsutsui, H., et al. (2008). Oxidative stress and mitochondrial DNA damage in heart failure. *Circulation Journal*, *72*, A31–A37. https://doi.org/10.1253/circj.cj-08-0014.

Vincent, J. L., et al. (2006). Sepsis in European intensive care units: Results of the SOAP study. *Critical Care Medicine*, *34*, 344–353. https://doi.org/10.1097/01.ccm.0000194725.48928.3a.

Wang, Q., & Zou, M. H. (2018). Measurement of reactive oxygen species (ROS) and mitochondrial ROS in AMPK knockout mice blood vessels. *Methods in Molecular Biology*, *1732*, 507–517. https://doi.org/10.1007/978-1-4939-7598-3_32.

Wu, J., et al. (2013). Activation of NLRP3 inflammasome in alveolar macrophages contributes to mechanical stretch-induced lung inflammation and injury. *Journal of Immunology*, *190*, 3590–3599. https://doi.org/10.4049/jimmunol.1200860.

Zhou, R., et al. (2011). A role for mitochondria in NLRP3 inflammasome activation. *Nature*, *469*, 221–225. https://doi.org/10.1038/nature09663.

CHAPTER FIVE

Quantification of intracellular and mitochondrial ATP content in macrophages during lipopolysaccharide-induced inflammatory response

Paulraj Kanmani[a] and Guochang Hu[a,b,*]

[a]Department of Anesthesiology, University of Illinois College of Medicine, Chicago, IL, United States
[b]Department of Pharmacology & Regenerative Medicine, University of Illinois College of Medicine, Chicago, IL, United States
*Corresponding author: e-mail address: gchu@uic.edu

Contents

Abstract

Sepsis, a condition characterized by systemic infection that becomes aggravated and dysregulated, is a significant cause of mortality in critically ill patients. Emerging evidence suggests that severe sepsis is often accompanied by alterations in cell

Methods in Cell Biology, Volume 194
ISSN 0091-679X
https://doi.org/10.1016/bs.mcb.2024.01.006

metabolism, particularly mitochondrial dysfunction, resulting in multiorgan failure. Normally, metabolically active cells or tissues exhibit higher levels of mitochondrial turnover, respiration, and adenosine triphosphate (ATP) synthesis. However, during sepsis, these processes become overwhelmed or dysregulated, leading to impaired ATP production in mitochondria. Here, we present two straightforward protocols for quantifying ATP production from mitochondria in bone marrow-derived macrophages (BMDMs). Our workflow facilitates the easy isolation of BMDMs and mitochondria from BMDMs treated with lipopolysaccharide (LPS), the major cell wall component of Gram-negative bacteria. We quantified intracellular and mitochondrial ATP production in macrophages *in vitro* using this protocol. The results indicated a decrease in mitochondrial ATP content in BMDMs in response to LPS. With minimal adjustments, this method can be adapted for use with various human and mouse primary cells and cell lines.

1. Introduction

Sepsis represents a pervasive systemic inflammatory response that arises from infections or trauma, posing a substantial and pressing public health concern (Fleischmann et al., 2016; Mu et al., 2020). In the United States, sepsis-related acute organ dysfunction is responsible for a rising number of deaths, with an estimated annual toll of approximately 270,000 (Magill et al., 2023; Paoli et al., 2018). Bacterial endotoxins, as well as viral and fungal molecular patterns, stimulate pathogenic signaling by interacting with specific receptors. This interaction leads to the secretion of inflammatory cytokines and mediators, ultimately causing tissue injury and dysfunction in vital organs (Hotchkiss et al., 2016; Huang et al., 2019; Rubio et al., 2019; Tamayo et al., 2011). Sepsis can potentially progress to septic shock, characterized by a dangerous drop in blood pressure, further exacerbating organ damage (Louis Vincent et al., 2019; Thompson et al., 2019).

Mitochondria, the vital organelles within cells, play a central role in regulating cellular metabolism (Park & Zmijewski, 2017). They are responsible for the utilization of cellular oxygen and the production of adenosine triphosphate (ATP), which serves as the primary energy source for normal cellular functions (Rolfe & Brown, 1997). ATP synthesis occurs through the electron transport chain, located in the inner mitochondrial membrane. This chain consists of four complexes (designated as I through IV) along with associated mobile electron carriers. The synthesis of ATP involves the conversion of adenosine diphosphate (ADP) to ATP through oxidative phosphorylation, which is coupled with the glycolytic pathway of the Krebs cycle (Park & Zmijewski, 2017). Glucose undergoes metabolism *via* the glycolytic pathway, leading to the formation of pyruvate. Within the mitochondria, pyruvate and fatty acids are further converted into acetyl-CoA, ultimately resulting in the generation of ATP through a series of oxidative phosphorylation (Park & Zmijewski, 2017; Singer, 2014). In the majority of cell types, including human cells, oxidative phosphorylation in the mitochondria is responsible for generating over 95% of the intracellular ATP required for cellular processes (Erecińska & Silver, 1989). Consequently, the intracellular ATP content serves as a reliable indicator for evaluating mitochondrial ATP production.

Abnormalities in mitochondrial function contribute to increased production of reactive oxygen species (ROS) and reduced ATP levels (Kozlov et al., 2011). Mitochondrial dysfunction is frequently associated with sepsis-related organ failure and patient outcomes (Brealey et al., 2002; Exline & Crouser, 2008; Lorente et al., 2012). Inflammatory factors or the inflammatory milieu, including reactive oxygen and nitrogen species, nitrogen oxide, and carbon monoxide, have been reported to impair components of the mitochondrial electron transport chain complexes and respiration (Larsen, 2012; Singer, 2007). Numerous studies have unequivocally demonstrated that sepsis significantly impairs cellular oxygen consumption, with mitochondria accounting for more than 90% of the oxygen utilized. The impaired oxygen consumption and associated mitochondrial dysfunction may contribute to the progression and severity of sepsis (Andrades et al., 2005; Brealey et al., 2004). Decreased components of electron transport chain complexes have been observed in a bacterial sepsis model with impaired oxidative metabolic rates (Haden et al., 2007). Mounting evidence strongly suggests that mitochondrial dysfunction associated with sepsis detrimentally impacts respiratory chain function, leading to a notable reduction in ATP production (Belikova et al., 2007; Brealey et al., 2002). Under normal conditions, mitochondrial respiration is more active in the presence of higher levels of oxygen and ADP (state 3 respiration), while it is less active at lower levels of ADP (state 4 respiration). Mitochondria extracted from both rat and human skeletal muscle during septic shock exhibit a substantial decline in state 3 respiration levels. This decline serves as a compelling indicator of impaired electron transport chain complexes and subsequent impairment in mitochondrial ATP production (Boveris, 2002). Higher levels of state 4 respiration, indicative of abnormal permeability of the inner and outer mitochondrial membranes, have been observed in the liver during acute endotoxemia (Crouser, 2004). Acute respiratory distress syndrome (ARDS) is a rapidly progressing condition that significantly contributes to the rising mortality rates among sepsis patients (Matthay et al., 2019). In sepsis patients, ARDS is accompanied by oxidative stress caused by excessive ROS production (Galley, 2011). Infection and the resulting inflammation exacerbate ROS production and mitochondrial damage, ultimately impairing mitochondrial ATP production, compromising the integrity of the epithelial barrier, and triggering apoptosis in lung epithelial cells within sepsis models (Bhatti et al., 2017; Carchman et al., 2013; Chang et al., 2015). In our recent study, we revealed that the deletion of p120 catenin in macrophages led to mitochondrial dysfunction, characterized by a decrease in mitochondrial membrane potential and a notable reduction in ATP production upon LPS stimulation (Kanmani et al., 2023). Overall, accumulating evidence suggests that the reduction of ATP production and mitochondrial dysfunction are key mechanisms in the pathogenesis of sepsis. Therefore, it is crucial to establish consistent and reproducible procedures for quantifying mitochondrial ATP production in different host systems. In this chapter, we introduce two simplified protocols for assessing mitochondrial ATP production *in vitro* specifically for macrophages. These versatile protocols can be utilized across a range of primary cells and cell lines derived from both humans and animals.

2. Materials

2.1 Common disposables

- 15 mL Falcon centrifuge tube; sterile (Denville, #C1071-p) (see *Note 1*)
- 50 mL Falcon centrifuge tube; sterile (Denville, #C1062-p) (see *Note 1*)
- 60 mm tissue culture dish (Denville, #T1116) (see *Notes 1 and 2*)
- 100 mm tissue culture dish (Denville, #T1110) (see *Notes 1 and 2*)
- 6-well tissue culture plate (Denville, #1020A) (see *Notes 1 and 2*)
- Cell strainer (70-μm pore size), sterile (Falcon, #351029) (see *Notes 1 and 3*)
- Syringe filter (0.2-μm, pore size; #4652, PALL Life Sciences) (see *Notes 1 and 3*)
- 10 mL syringe (BD, #303134) (see *Note 4*)
- 25-gauge needle, sterile (BD, #305122) (see *Note 4*)

2.2 Surgical accessories

- Sharp-pointed dissecting scissors (Fisherbrand, #08-940) (see *Note 1*)
- Delicate dissecting scissors (Fisherbrand, #08-951-5) (see *Note 1*)

- Sterile standard scalpels (Fisherbrand, #12-460-451) (see *Note 1*)
- Straight serrated medium point forceps (Fisherbrand, #16-100-109) (see *Note 1*)
- Curved medium point general purpose forceps (Fisherbrand, #16-100-110) (see *Note 1*)
- General purpose tweezers with straight tips (Fisherbrand, #17-467-235) (see *Note 1*)

2.3 Cells and reagents

- L929 cells; NCTC clone 929 Areolar fibroblasts; Mouse (CCL-1, ATCC, Rockville, MD) (see *Note 5*)
- BMDMs isolated from C57BL/6J mice (Jackson Laboratory, USA) (see *Notes 5 and 6*)
- Roswell Park Memorial Institute (RPMI) 1640 medium (Gibco, #11875-093) (see *Notes 1 and 7*), supplemented with 10% fetal bovine serum (FBS; Gibco, #26140079); 100 U mL^{-1} penicillin and 100 μg mL^{-1} streptomycin (10,000 U mL^{-1}, Gibco, #15140122) (see *Notes 1 and 8*)
- Dulbecco's modified Eagle's medium (DMEM; Gibco, #11965-092) supplemented with 10% fetal bovine serum; 100 U mL^{-1} penicillin and 100 μg mL^{-1} streptomycin (see *Notes 1, 7 and 8*)
- Phosphate buffered saline (PBS; Corning, #21-040-CV) (see *Note 1*)
- Lipopolysaccharide (LPS; InvivoGen, #NC0202558) (see *Notes 1 and 9*)
- 70% ethanol (70 mL of ethanol mixed with 30 mL of water) (see *Notes 1 and 10*)
- 2 mL Dounce homogenizer (Fisher scientific, #50-194-5204)
- ATP assay kit (Abcam, #Ab83355) (see *Notes 1 and 11*)
- Mitochondria isolation kit for cultured cells (Thermo Scientific, #89874) (see *Notes 1 and 11*)
- Potassium chloride (Sigma, #7447-40-7)
- Mannitol (Sigma, #BP1007)
- Sucrose (Sigma, #S0389)
- Magnesium chloride (Sigma, #M8266)
- Dithiothreitol (RPI, #D11000-100.0)
- HEPES (Sigma, #H3375)
- EDTA (Sigma, #E9884)
- EGTA (Sigma, #324626)
- Non-enzymatic cell dissociation solution (Corning, #27420004) (see *Note 1*)

- Homogenization buffer (10 mM KCl, 5 mM HEPES, 1.5 mM $MgCl_2$, 20 mM mannitol, 70 mM sucrose, 1 mM EDTA, 1 mM EGTA, 1.5 mM dithiothreitol)

2.4 Equipment

- Laboratory biosafety cabinet (SterilGard Hood, USA) (see *Note 1*)
- Humidified cell culture incubator sets at 37 °C with 5% CO_2 (Fischer Scientific, Model: 3530) (see *Note 1*)
- Enzyme-linked immunosorbent assay (ELISA) reader (Bio-Rad, USA) (see *Note 1*)
- Standard bench top centrifuge (Eppendorf, Centrifuge 5415R) (see *Note 1*)

2.5 Software

- Prism software (see *Notes 1 and 12*)

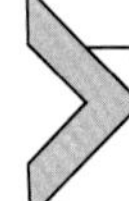

3. Methods

3.1 Preparation of L929 conditional medium

1. Seed L929 cells (5×10^4 cells mL^{-1}) in a cell culture plate containing DMEM complete culture medium and culture at 37 °C, 5% for 3 days (see *Notes 5 and 13*)
2. Collect the cell-free supernatant from the cultured plate, filter with a syringe filter (0.2 μm pore size) and store at −80 °C (see *Notes 13 and 14*)
3. Prepare the differentiation medium by mixing L929 conditional medium (30%) and RPMI complete medium (70%) for culturing BMDMs (see *Notes 5 and 15*)

3.2 Extraction and culture of murine bone marrow-derived macrophages (BMDMs)

1. Euthanize 6- to 8-week-old C57BL/6 mice by intraperitoneal injection of 5 mg kg^{-1} xylazine and 100 mg kg^{-1} ketamine, followed by cervical dislocation. The procedures were approved by the Institutional Animal Care and Use Committee of the University of Illinois at Chicago (see *Notes 6 and 16*)
2. Sterilize the legs (femurs and tibias), peel the skin, remove the tissues from each leg, cut off the legs at hip joint, and place them in a 60 mm dish containing ice cold PBS (see *Note 17*)

3. Take the dish into the cell culture hood, remove the extra tissues around the legs, and put them in a 60 mm dish containing 70% ethanol for 2 min (see *Note 17*)
4. Transfer the sterilized and clean femurs and tibias into a fresh 60 mm dish containing RPMI complete medium
5. Hold the bones with sterile tweezers and cut out the femurs and tibias at both ends using a sharp sterile scissor to expose the bone marrow (see *Note 18*)
6. Aspirate the sterile RPMI medium into the sterile 10 mL syringe fit with 25G needle and put the needle into the bone lumen and flush out the bone marrow into a fresh 60 mm dish containing RPMI complete medium until the bones become white visibly (see *Note 19*)
7. Aspirate the harvested bone marrow using a 10 mL syringe and pass through the 25G needle to dissociate cells (see *Notes 4 and 20*)
8. Transfer the cell suspension in RPMI medium into a fresh 15 mL Falcon tube by passing through the 70-μm pore size cell strainer and centrifuge at $300 \times g$ for 5 min at room temperature (see *Notes 4 and 21*)
9. Discard the supernatant and resuspend the pellet with 1 mL of red cell lysis buffer to lyse red blood cells (RBCs). Pipette up and down for 30–40 s and immediately add 15 mL of PBS and centrifuge at $300 \times g$ for 5 min at room temperature (see *Notes 1 and 22*)
10. Discard the supernatant and repeat the lysis steps until the complete lysis of RBCs. Finally, resuspend cell pellet in 20 mL differentiation medium and plated cells on 100 mm tissue culture plate (see *Note 23*)
11. Keep the plate in the humidified incubator at 37 °C, 5% CO_2 for 90 min to remove stromal cells, collect the non-adherent cells (4×10^5 cells) into a fresh dish containing 10 mL of differentiation medium, and incubate cells in a humidified incubator at 37 °C, 5% CO_2 (see *Note 24*)
12. Add fresh 5 mL of differentiation medium into the culture dish and continue the incubation at the same condition as above on day 3 and 5 of culture (see *Note 25*)
13. Remove the old medium and add 10 mL of fresh differentiation medium into the culture dish and incubate at 37 °C, 5% CO_2 on day 6 of culture
14. On day 7 of culture, bone marrow precursors are fully differentiated into mature macrophages that were used for the analysis of mitochondrial ATP production *in vitro* (see *Note 26*)

3.3 Measurement of intracellular ATP

1. Measure intracellular ATP in the cells according to the manufacturer's protocol with slight modification
2. Seed BMDMs (1×10^6) into a 6-well plate and treat with LPS ($20\,ng\,mL^{-1}$) for 4 h (see *Notes 26 and 27*)
3. Wash the cells with PBS and resuspend with 100 μL of ATP assay buffer
4. Homogenize the cells by pipetting up and down for 1 min and centrifuge at $13{,}000 \times g$, 4 °C for 5 min (see *Note 28*)
5. Collect the cell-free supernatant into fresh Eppendorf and keep on ice
6. Add ice cold 4 M perchloric acid to a final concentration of 1 M in the cell culture supernatant for deproteinization, vortex to mix well, and keep on ice for 5 min (see *Note 29*)
7. Centrifuge at $13{,}000 \times g$, 4 °C for 2 min and transfer the supernatant into a fresh tube
8. Precipitate extra perchloric acid in the supernatants by adding ice cold 2 M potassium hydroxide (KOH) that equals to 35% of the total volume of samples (pH should be at 7)
9. Centrifuge at $13{,}000 \times g$, 4 °C for 15 min and collect the deproteinized samples into a fresh tube to perform ATP assay
10. Prepare ATP standards with ATP assay buffer to generate 0–10 nM/well
11. Mix 50 μL of ATP standard or deproteinized samples with the 50 μL of reaction mixer, incubate at room temperature for 30 min, protect from light (see *Note 30*), and measure using a microplate reader at optical density (OD) 570 nm

3.4 Isolation of mitochondria from BMDM cells

Method I

1. Seed BMDMs (2×10^7) in a 60 mm cell culture plate and treat with LPS ($20\,ng\,mL^{-1}$) for 4 h
2. Wash cells with PBS and add non-enzymatic cell dissociation solution into the wells to dissociate cells from the plate, transfer to fresh Eppendorf tubes, and centrifuge at $850 \times g$ for 2 min (see *Note 31*)
3. Wash the cell pellets with PBS and centrifuge at $850 \times g$ for 2 min
4. Aspirate the PBS and mix cell pellets with four volumes of ice-cold homogenizing buffer and physically shear the cells (20–25 up and down passes) using a cold Dounce homogenizer, and keep on ice for 2 min (see *Note 32*)
5. Centrifuge the homogenized cell solution at $800 \times g$ for 5 min

6. Transfer the resulting supernatant to a fresh tube and cell pellets mix with homogenizing buffer and repeat the steps 4–6 (see *Note 32*)
7. Pool all the supernatants and centrifuge at 12,000 × *g* for 15 min at 4 °C (see *Note 33*)

Methods II

1. Isolate the mitochondria from the BMDMs according to a manufacture's kit protocol with appropriate modification
2. Resuspend LPS treated cell pellets with 800 μL of mitochondrial isolation reagent A, vortex for 5 s, and keep on ice for 2 min
3. Add 10 μL of mitochondrial isolation reagent B into the same sample tube, vortex at maximum speed for 5 min, incubate on ice for 5 min, and vortex at maximum speed every minute
4. Add 800 μL of mitochondrial isolation reagent C into the same sample tubes, invert for 5–6 times to mix well, and centrifuge at 700 × *g* for 10 min at 4 °C
5. Transfer the upper supernatant to a fresh tube and centrifuge at 12,000 × *g* for 15 min at 4 °C (see *Note 33*)
6. Mix the pellets containing mitochondria with 500 μL of mitochondrial isolation reagent C and centrifuge at 12,000 ×*g* for 5 min at 4 °C

3.5 Measurement of mitochondrial ATP

1. Wash the mitochondria pellets with PBS and centrifuge at 12,000 × *g* for 5 min at 4 °C
2. Aspirate the PBS and resuspend the pellets with 100 μL of ATP assay buffer
3. Homogenize the mitochondria by gently pipetting up and down for 1 min and then centrifuge at 13,000 × *g*, 4 °C for 5 min (see *Note 28*)
4. Repeat steps 4–10 in Section 3.3

3.6 Data analysis

1. The fully differentiated BMDMs were treated with LPS, and the total lysate and mitochondrial samples were used for quantification of intracellular and mitochondrial ATP content *in vitro* and the data are shown in Fig. 1A and B
2. The prism software was used for analyzing experimental data and plotting bar/scattering graphs (see *Note 12*)
3. Paired Student's *t*-test was employed to assess statistically significant difference between the experimental groups

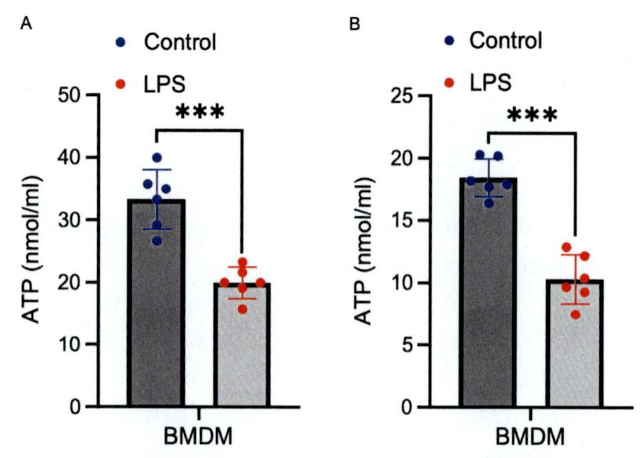

Fig. 1 Measurement of intracellular and mitochondrial ATP content in BMDMs following lipopolysaccharide stimulation. The mature BMDMs isolated from C57BL/6J mice were treated with lipopolysaccharide (LPS, $20\,ng\,mL^{-1}$) for 4 h, as detailed in Sections 3.2 and 3.3. (A) Measurement of intracellular ATP content. The cell lysates were used for the quantification of intracellular ATP content as detailed in Section 3.3. (B) Measurement of mitochondrial ATP content. Mitochondria was isolated from the mature BMDMs treated with LPS ($20\,ng\,mL^{-1}$) for 4 h, as detailed in Sections 3.2–3.4. Freshly isolated mitochondria were lysed immediately for the measurement of mitochondrial ATP content as detailed in Sections 3.3 and 3.4. The data represented as mean ± SD. $^{***}P < 0.001$ (Paired Student's *t*-test) (see Ref. Kanmani et al., 2023).

4. Notes

1. The providers and catalog numbers are provided as reference, but equivalent products can be purchased from a variety of sources.
2. Smaller and larger cell culture dishes or plates can be employed depending on the appropriate number of cells.
3. Sterile cell strainers and syringe filters should be used and can be purchased from different companies.
4. Alternative sizes of syringes and needles can be utilized as substitutes for those specified in Section 2.1.
5. L929 cells are non-hazardous and can be employed for producing macrophage colony-stimulating factor. However, commercially available macrophage colony-stimulating factor can also be utilized for differentiating BMDMs as an alternative approach.
6. Ensure that all procedures are conducted in a sterile environment, such as a biosafety cabinet, during the isolation of BMDMs from C57BL mice. Alternatively, BMDMs can be isolated from other mouse strains.

7. BMDMs, RPMI, DMEM, FBS, and PBS are classified as non-hazardous, but it is important to handle them using appropriate certified personal protective equipment (PPE) as a precautionary measure.
8. When handling penicillin and streptomycin, it is essential to wear suitable certified PPE due to potential risks, including allergic skin reactions, allergy, or asthma symptoms, and breathing difficulties if inhaled, as well as potential harm to fertility and the unborn child.
9. Lipopolysaccharide should be manipulated while wearing appropriate certified PPE to prevent the induction of inflammation. It can be substituted with other types of inflammatory agents if desired.
10. Ethanol must be handled while wearing appropriate certified PPE and kept away from open flames, heat sources, and sparks. It should be stored in dedicated cabinets for flammable substances. Ethanol is highly flammable in both liquid and vapor states and can cause severe eye irritation.
11. Isolation and assay kits can be obtained from various manufacturers; however, it is important to follow the manufacturer's protocol for successful isolation and ATP assay.
12. The software version and provider mentioned are provided as references, but equivalent software options can be purchased from different providers (e.g., ModFit LT™, Cyflogic).
13. Cell culture steps must be carried out within a Class II Biological Safety Cabinet to prevent bacterial and fungal contamination. Cells should be grown in DMEM medium, allowing it to reach normal temperature before use.
14. The conditioned medium containing colony-stimulating factor should be filtered to ensure complete elimination of L929 cells and promptly stored at −80 °C until ready for experimental use.
15. The macrophage differentiation medium should be prepared by mixing conditional and RPMI medium at a ratio of 30:70. It should be stored at 4 °C, with a maximum storage period of 1 month for further use.
16. While the age of the mice is not a significant factor, younger mice are generally preferable to older ones. Once euthanized, the mice should be used immediately.
17. It is advisable to remove the blood before cutting off the legs. Simply open the abdomen, cut the portal vein, and use a tissue towel to absorb the blood.
18. Euthanizing and cutting off the legs can be performed on the bench, but the subsequent isolation steps should be carried out within a biosafety

cabinet. A 100 mm dish can also be used for these steps. The bones should be soaked in 70% ethanol for no more than 1–2 min. Ensure all materials are sterilized before use to prevent contamination.

19. The bone should be held firmly with tweezers and the needle should be gently inserted into the middle part of the bone lumen. The medium is pushed to flush out the bone marrow. This step should be repeated at least 2–3 times to completely flush out the bone marrow.
20. Different sizes of syringes and needles can be used to aspirate the bone marrow-containing medium and push it through the needle. This step should be repeated at least 2–3 times to ensure complete dissociation of the bone marrow or cells.
21. The cell suspension should be filtered through a cell strainer to remove and prevent the mixing of tissues and broken bones with the cells.
22. Prior to seeding the cells in the cell culture plate, red blood cells (RBCs) in the cell pellet should be lysed. The lysis step can be repeated until complete lysis of RBCs. A commercially available RBC lysis buffer, such as 1× RBC lysis buffer (#00-4333-57, Invitrogen), can be used.
23. There is no specific volume of differentiation medium, and less than 20 mL of medium can be used to resuspend the cells. After counting the number of cells in the medium by mixing a sample with trypan blue at a 1:1 ratio, the cells are distributed into different cell culture plates or flasks.
24. The stromal cells should be removed from the cell suspension and only non-adherent BMDM cells are transferred to fresh cell culture plates. At least 10 mL of differentiation medium should be used for culturing the cells.
25. Fresh medium should be added into the cell culture to ensure that the volume is half of the original volume of medium initially added to the cell culture plate. These steps are crucial for the differentiation of cells into mature macrophages.
26. The mature cells firmly attached to the plates should be removed by adding cell dissociation buffer and then seeding an appropriate volume of cells in a new 6-well plate or 60 mm dish for experiments.
27. Lipopolysaccharide can be purchased from various companies. The optimal concentration and duration of lipopolysaccharide treatment for this study and BMDMs have been determined. However, the concentration or duration of lipopolysaccharide treatment may vary depending on the cell type and mouse strain.

28. Cells can be homogenized by pipetting or using a Dounce homogenizer. Most of the steps should be performed on ice to maintain sample integrity.

29. Deproteinization can be achieved by adding perchloric acid or by using a deproteinizing sample preparation kit (e.g., #ab204708, Abcam).

30. The ATP standard should be prepared according to the manufacturer's instructions. Mixed samples or standards should be stored in a dark place, or the assay plate can be wrapped in aluminum foil. The samples must be measured immediately on a microplate reader, although the manufacturer suggests that the reaction may remain stable for up to 2 h.

31. Non-enzymatic cell dissociation solution is recommended to avoid cell rupture or damage.

32. The cells should be homogenized by passing them 20–25 times in a sterile glass Dounce. It is not advisable to perform more than 30 up-down strokes with the pestle. Repeat the homogenization process 1–2 times after transferring the supernatant to a fresh tube without touching the pellets.

33. To obtain a more purified fraction of mitochondria with a reduction of lysosomal and peroxisomal contaminants by over 50%, centrifuge at $3000 \times g$ for 15 min.

5. Concluding remarks

Mitochondria serve as the primary site within cells for the production of high-energy ATP through oxidative phosphorylation. In the context of sepsis, there is a notable decrease in mitochondrial oxygen consumption and reduced activity of complex I, resulting in a decline in ATP synthesis and overall ATP content (Arulkumaran et al., 2016; Vanasco et al., 2012). Among critically ill patients, downregulation of respiratory protein subunits and transcripts for complexes I and IV has been observed, and such patients who do not survive demonstrate a more prolonged recovery process and experience a greater reduction in ATP levels (Brealey et al., 2002; Carre et al., 2010). Collectively, research conducted on animals and septic patients highlights the significant role of mitochondrial ATP production in the pathophysiology and progression of sepsis. The presented protocols provide simple and straightforward techniques for quantifying intracellular and mitochondrial ATP content in cells, utilizing a reliable colorimetric method. The ATP assay protocol relies on the phosphorylation of glycerol to produce a measurable product, which can be conveniently quantified using colorimetric or fluorometric methods. The employed approach for isolating bone marrow precursors has the potential to generate an ample supply of mature BMDMs suitable for accurate analysis of ATP content. However, it is crucial to exercise caution and employ the cell isolation technique meticulously to ensure the acquisition of pure and abundant cells, enabling precise measurement of intracellular and mitochondrial ATP. In our study, we assessed both the intracellular ATP content and mitochondrial ATP production in BMDMs upon LPS stimulation. This ATP assay can also be applied to quantify ATP content in various tissue extracts, cell lysates, and biological fluids, broadening its applicability beyond our specific experimental setting.

Acknowledgments

The work was supported by an R01 grant (#HL152696, PI: G.H.) and an R01 grant (#HL104092, PI: G.H.) from the National Heart, Lung, and Blood Institute of the National Institutes of Health, and an R21 grant (#AI152249, PI: G.H.) from the National Institute of Allergy and Infectious Diseases of the National Institutes of Health.

Competing interests

The authors have no conflicts of interest to declare.

References

Andrades, M., et al. (2005). Oxidative parameters differences during non-lethal and lethal sepsis development. *Journal of Surgical Research*, *125*, 68–72. https://doi.org/10.1016/j.jss.2004.11.008.

Arulkumaran, N., et al. (2016). Mitochondrial function in sepsis. *Shock*, *45*, 271–281. https://doi.org/10.1097/SHK.0000000000000463.

Belikova, I., et al. (2007). Oxygen consumption of human peripheral blood mononuclear cells in severe human sepsis. *Critical Care Medicine*, *35*, 2702–2708. https://doi.org/10.1097/01.ccm.0000295593.25106.c4.

Bhatti, J. S., et al. (2017). Mitochondrial dysfunction and oxidative stress in metabolic disorders—A step towards mitochondria based therapeutic strategies. *Biochimica et Biophysica Acta. Molecular Basis of Disease*, *1863*, 1066–1077. https://doi.org/10.1016/j.bbadis.2016.11.010.

Boveris, A. (2002). The role of mitochondrial nitric oxide synthase in inflammation and septic shock. *Free Radical Biology and Medicine*, *33*, 1186–1193. https://doi.org/10.1016/s0891-5849(02)01009-2.

Brealey, D., et al. (2002). Association between mitochondrial dysfunction and severity and outcome of septic shock. *Lancet*, *360*, 219–223. https://doi.org/10.1016/S0140-6736(02)09459-X.

Brealey, D., et al. (2004). Mitochondrial dysfunction in a long-term rodent model of sepsis and organ failure. *American Journal of Physiology. Regular, Integrative and Comparative Physiology*, *286*, R491–R497. https://doi.org/10.1152/ajpregu.00432.2003.

Carchman, E. H., et al. (2013). Experimental sepsis-induced mitochondrial biogenesis is dependent on autophagy, TLR4, and TLR9 signaling in liver. *FASEB Journal, 27*, 4703–4711. https://doi.org/10.1096/fj.13-229476.

Carre, J. E., et al. (2010). Survival in critical illness is associated with early activation of mitochondrial biogenesis. *American Journal of Respiratory and Critical Care Medicine, 182*, 745–751. https://doi.org/10.1164/rccm.201003-0326OC.

Chang, A. L., et al. (2015). Redox regulation of mitophagy in the lung during murine *Staphylococcus aureus* sepsis. *Free Radical Biology and Medicine, 78*, 179–189. https://doi.org/10.1016/j.freeradbiomed.2014.10.582.

Crouser, E. D. (2004). Abnormal permeability of inner and outer mitochondrial membranes contributes independently to mitochondrial dysfunction in the liver during acute endotoxemia. *Critical Care Medicine, 32*, 478–488. https://doi.org/10.1097/01.CCM.0000109449.99160.81.

Erecińska, M., & Silver, I. A. (1989). ATP and brain function. *Journal of Cerebral Blood Flow & Metabolism, 9*, 2–19.

Exline, M. C., & Crouser, E. D. (2008). Mitochondrial mechanisms of sepsis-induced organ failure. *Frontiers in Bioscience, 13*, 5030–5041. https://doi.org/10.2741/3061.

Fleischmann, C., et al. (2016). Assessment of global incidence and mortality of hospital-treated sepsis. Current estimates and limitations. *American Journal of Respiratory and Critical Care Medicine, 193*, 259–272. https://doi.org/10.1164/rccm.201504-0781OC.

Galley, H. F. (2011). Oxidative stress and mitochondrial dysfunction in sepsis. *British Journal of Anesthesiology, 107*, 57–64. https://doi.org/10.1093/bja/aer093.

Haden, D. W., et al. (2007). Mitochondrial biogenesis restores oxidative metabolism during *Staphylococcus aureus* sepsis. *American Journal of Respiratory and Critical Care Medicine, 176*, 768–777. https://doi.org/10.1164/rccm.200701-161OC.

Hotchkiss, R. S., et al. (2016). Sepsis and septic shock. *Nature Reviews. Disease Primers, 2*, 16045. https://doi.org/10.1038/nrdp.2016.45.

Huang, M., et al. (2019). The pathogenesis of sepsis and potential therapeutic targets. *International Journal of Molecular Sciences, 20*, 5376. https://doi.org/10.3390/ijms20215376.

Kanmani, P., et al. (2023). p120-Catenin suppresses NLRP3 inflammasome activation in macrophages. *American Journal of Physiology. Lung Cellular and Molecular Physiology, 324*, L596–L608. https://doi.org/10.1152/ajplung.00328.2022.

Kozlov, A. V., et al. (2011). Mitochondrial dysfunction and biogenesis: Do ICU patients die from mitochondrial failure? *Annals of Intensive Care, 1*, 41. https://doi.org/10.1186/2110-5820-1-41.

Larsen, F. J. (2012). Regulation of mitochondrial function and energetics by reactive nitrogen oxides. *Free Radical Biology and Medicine, 53*, 1919–1928. https://doi.org/10.1016/j.freeradbiomed.2012.08.580.

Lorente, L., et al. (2012). Survival and mitochondrial function in septic patients according to mitochondrial DNA haplogroup. *Critical Care, 16*, R10. https://doi.org/10.1186/cc11150.

Louis Vincent, J., et al. (2019). Frequency and mortality of septic shock in Europe and North America: A systematic review and meta-analysis. *Critical Care, 23*, 196. https://doi.org/10.1186/s13054-019-2478-6.

Magill, S. S., et al. (2023). Epidemiology of sepsis in US children and young adults. *Open Forum Infectious Diseases, 10*, ofad218. https://doi.org/10.1093/ofid/ofad218.

Matthay, M. A., et al. (2019). Acute respiratory distress syndrome. *Nature Reviews. Disease Primers, 5*, 18. https://doi.org/10.1038/s41572-019-0069-0.

Mu, X., et al. (2020). Identification of a novel antisepsis pathway: Sectm1a enhances macrophage phagocytosis of bacteria through activating GITR. *The Journal of Immunology, 205*, 1633–1643. https://doi.org/10.4049/jimmunol.2000440.

Paoli, C. J., et al. (2018). Epidemiology and costs of sepsis in the United States—An analysis based on timing of diagnosis and severity level. *Critical Care Medicine*, *46*, 1889–1897. https://doi.org/10.1097/CCM.0000000000003342.

Park, D. W., & Zmijewski, J. W. (2017). Mitochondrial dysfunction and immune cell metabolism in sepsis. *Infection & Chemotherapy*, *49*, 10–21. https://doi.org/10.3947/ic.2017.49.1.10.

Rolfe, D. F., & Brown, G. C. (1997). Cellular energy utilization and molecular origin of standard metabolic rate in mammals. *Physiological Reviews*, 77, 731–758. https://doi.org/10.1152/physrev.1997.77.3.731.

Rubio, I., et al. (2019). Current gaps in sepsis immunology: New opportunities for translational research. *The Lancet Infectious Diseases*, *19*, e422–e436. https://doi.org/10.1016/S1473-3099(19)30567-5.

Singer, M. (2007). Mitochondrial function in sepsis: Acute phase versus multiple organ failure. *Critical Care Medicine*, *35*, S441–S448. https://doi.org/10.1097/01.CCM.0000278049.48333.78.

Singer, M. (2014). The role of mitochondrial dysfunction in sepsis-induced multi-organ failure. *Virulence*, *5*, 66–72. https://doi.org/10.4161/viru.26907.

Tamayo, E., et al. (2011). Pro- and anti-inflammatory responses are regulated simultaneously from the first moments of septic shock. *European Cytokine Network*, *22*, 82–87. https://doi.org/10.1684/ecn.2011.0281.

Thompson, K., et al. (2019). Sepsis and septic shock: Current approaches to management. *Internal Medicine Journal*, *48*, 160–170. https://doi.org/10.1111/imj.14199.

Vanasco, V., et al. (2012). Endotoxemia impairs heart mitochondrial function by decreasing electron transfer, ATP synthesis and ATP content without affecting membrane potential. *Journal of Bioenergetics and Biomembranes*, *44*, 243–252. https://doi.org/10.1007/s10863-012-9426-3.

CHAPTER SIX

Analysis of cytosolic mtDNA release during *Staphylococcus aureus* infection

Caterina Licini[a], Gloria D'Achille[b], Nada Dhaouadi[a], Ilaria Nunzi[a], Fabio Marcheggiani[a], Matteo Fabbri[c], Monica Mattioli-Belmonte[a,d], Gianluca Morroni[b], and Saverio Marchi[a,d,*]

[a]Department of Clinical and Molecular Sciences, Marche Polytechnic University, Ancona, Italy
[b]Microbiology Unit, Department of Biomedical Sciences and Public Health, Marche Polytechnic University, Ancona, Italy
[c]Section of Legal Medicine, Department of Translational Medicine, University of Ferrara, Ferrara, Italy
[d]Advanced Technology Center for Aging Research, IRCCS INRCA, Ancona, Italy
*Corresponding author: e-mail address: s.marchi@staff.univpm.it

Contents

Abstract

Methicillin-resistant *Staphylococcus aureus* (MRSA) is one of the principal human pathogens, causing severe infections in skin wounds. MRSA infection triggers a cell response mainly by mitochondrial-mediated pathway, resulting in mitochondrial outer membrane permeabilization, extrusion of the mitochondrial inner membrane into the cytoplasm, and then spillage of mitochondrial DNA (mtDNA) into the cytoplasm.

Methods in Cell Biology, Volume 194
ISSN 0091-679X
https://doi.org/10.1016/bs.mcb.2024.09.003

The cell recognizes the discharged cytosolic mtDNA (cmtDNA) as "not-itself" because of mtDNA properties and triggers cascade events, such as the activation of inflammasomes. Here, we detail a method to detect and measure the mtDNA release into the cytoplasm in immortalized keratinocytes (HaCaT cells), after the infection with MRSA at different time points after the infection.

1. Introduction

Staphylococcus aureus is one of the main human pathogens, able to cause a wide range of diseases, from mild skin and wound infections to fatal sepsis and multiple organ failure. *S. aureus* colonized approximately 20–30% of healthy individuals, becoming a risk factor for subsequent infections. (Howden et al., 2023).

Besides its ability to evade the immune response and to acquire multiple antibiotic resistance traits, *S. aureus* is also able to induce host cell death through different strategies, including intracellular invasion, secretion of several toxins (*i.e.*, alpha-toxin and Panton-Valentine leukocidin), and extracellular vesicles (Zhang, Hu, & Rao, 2017) (Soe, Bedoui, Stinear, & Hachani, 2021) (Missiakas & Winstel, 2021) (Chen et al., 2022).

Host cell apoptosis induced by *S. aureus* occurs through various pathways, among which the mitochondrial-mediated pathway is the main one. Initiation of the pathway encompasses the BCL-2-associated X (BAX) and BCL-2 homologous antagonist/killer (BAK) activation, creating pores in the mitochondrial outer membrane, with consequent mitochondrial outer membrane permeabilization (MOMP) (Cosentino et al., 2022; McArthur et al., 2018; Wang et al., 2023).

The MOMP is often essential to trigger apoptosis (Guilbaud & Galluzzi, 2022; Vringer et al., 2024). This state causes the release of mitochondrial intermembrane space proteins into the cytoplasm, activating caspase proteases, and leads to cell death. (Riley et al., 2018; Riley & Tait, 2020) Furthermore, the extrusion of the mitochondrial inner membrane (MIM) into the cytoplasm is a BAX/BCL-2 antagonist/killer 1 (BAK)-mediated MOMP, and its permeabilization manages the spillage of mitochondrial DNA (mtDNA) into the cytoplasm. (Marchi, Guilbaud, Tait, Yamazaki, & Galluzzi, 2023; Riley et al., 2018).

During the pathogen infection, the host cell is capable of detecting *S. aureus* infection by pattern recognition receptors (PRRs), by which the cell identifies pathogen-associated molecular patterns (PAMPs) and/or damage-associated molecular patterns (DAMPs). Both PAMPs and DAMPs are known to cause pro-inflammatory cascades and/or trigger the immune response. (Marchi et al., 2023; Marchi, Morroni, Pinton, & Galluzzi, 2022; Riley & Tait, 2020).

What the cell recognizes as "not-itself," such as viral and/or bacterial proteins or DNA, can be considered as DAMP, as well as the cytosolic mtDNA (cmtDNA). Indeed, mtDNA is a round double-stranded DNA, with an independent genetic code and histones-free, recognized as "foreign" because of its peculiarities different from "self" DNA in cells. (Kim, Kim, & Chung, 2023; Riley & Tait, 2020).

One of the downstream mechanisms of the pathogen infection is the engagement of cyclic GMP-AMP synthase (cGAS), a nuclear and cytosolic protein that responds to cytosolic double-stranded DNA (dsDNA) molecules and catalyzes the formation of cyclic GMP–AMP (cGAMP) and able to stimulate an inflammatory response. Therefore, the cGAS-STING pathway is activated by not only exogenous DNA derived from pathogens but also mtDNA. (Kim et al., 2023; Marchi et al., 2023; Riley & Tait, 2020).

An alternative way by which the cell reacts to the pathogens and trigs by the presence of mtDNA in the cytosol is represented by activation of inflammasomes, and consequent stimulation of caspase 1, which lead to the processing of pro-Interleukin-1β (IL-1β) and pro-IL-18 into their mature form, following their secretion. (Carlos et al., 2017; Zhang et al., 2022) Otherwise, inflammasomes and caspase activity can operate upstream of the mtDNA release, causing mitochondrial damage and prompting the MOMP. (Akbal et al., 2022; Marchi et al., 2023; Riley & Tait, 2020).

Here, we detail a method to detect the mtDNA release into the cytoplasm, subsequent to the infection with methicillin-resistant *S. aureus* (MRSA) at 2, 4, and 6h upon the infection. We used HaCaT cells as a keratinocyte model for infection, as MRSA is known to cause severe chronic wounds in the skin (Pastar et al., 2014; Worster, Zawora, & Hsieh, 2015), due, at least in part, to mitochondrial damage and aberrant activation of the NLR family pyrin domain containing 3 (NLRP3) inflammasome. (Licini et al., 2024) To present the alteration in cmtDNA content in infected keratinocytes, in this paper we measured the cmtDNA by the evaluation of two mtDNA genes, mtND4, and mtDLoop, normalizing the results compared to the mtDNA in whole cells.

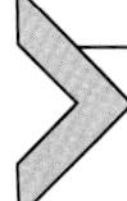

2. Materials

2.1 Common disposable (see Notes 4.1 and 4.2)

- 1–10 (613-0334), 2–200 (612-5755), and 100–1000 (613-1050) μL pipette tips (Avantor, Radnor, PA, USA)

- 1–10 (732-1486), 2–20 (732-1488), 2–100 (732-2385), 2–200 (732-1489), and 100–1000 (732-1491) μL sterile filtered pipette tips (Avantor)
- Flasks T-75 CytoOne®, TC treated, vented (CC7682-4875; Starlab, Milan, Italy)
- 2, 5, 10, 25 mL serological pipettes (E4860-0002, E4860-0005, E4860-0010, E4860-0025; Starlab)
- 1.5 mL microcentrifuge tubes (525-1126; Avantor)
- 15 and 50 mL tubes (E1415-0100, E1450-0100; Starlab)
- Cell scraper (734-2603; Avantor)
- 6-well plates, TC treated (734-2323; Avantor)
- MicroAmp™ Optical 8-Cap Strips (4323032; Applied Biosystems™, Waltham, MA, USA)
- Sterile Petri dishes 90 mm (101VIRR; Thermofisher Scientific, Waltham, MA, USA)
- Disposable sterile loops 10 μL (96701; Liofilchem, Roseto degli Abruzzi, Italy)
- Semi-micro polypropylene cuvettes (BRND759015; Avantor)

2.2 Cells and reagents

- HaCaT cells (CVCL_0038; Cellosaurus database)
- Dulbecco's Modification of Eagle's Medium supplemented with 4.5 g/L glucose (HG-DMEM), glutagro™, sodium pyruvate, phenol red (10–101-CV; Corning, Corning, NY, USA) (**Notes 4.1 and 4.3**)
- Fetal Bovine Serum (FBS) South America Superior (YOURSIAL-FBS-SA; Sial, Rome, Italy) (*see* **Notes 4.1 and 4.3**)
- Penicillin/Streptomycin 100× (ECB3001D; Euroclone, Pero, Italy) (NOTE 4.1 and 4.3)
- Dulbecco's Phosphate-Buffered Saline (D-PBS), 1× without calcium and magnesium (21-031-CV; Corning) (*see* **Notes 4.1 and 4.3**)
- Trypsin- EDTA solution 10× (T4174; Sigma-Aldrich, St. Louis, MO, USA) (*see* **Notes 4.1 and 4.3**)
- Trypan Blue stain 0.4% (T10282; Invitrogen™, Waltham, MA, USA) (*see* **Note 4.1**)
- 26 G Needle sterile Syringes (329652; East Rutherford, NJ, USA) (*see* **Note 4.1**)
- D-Mannitol (63565; Fluka Chemika) (*see* **Note 4.1**)
- Sucrose (S0389; Sigma-Aldrich) (*see* **Note 4.1**)
- HEPES (H3375; Sigma-Aldrich) (*see* **Note 4.1**)

- Potassium hydroxide (KOH) (221,473; Sigma-Aldrich) (*see* **Notes 4.1 and 4.4**)
- 3-Hydroxy-4-methoxyphenylacetic acid (EDTA) (716,391; Sigma-Aldrich) (*see* **Notes 4.1 and 4.5**)
- Bovine Serum Albumin (BSA) (A3059; Sigma-Aldrich)
- Protease inhibitors (s8820; Sigma-Aldrich) (*see* **Notes 4.1 and 4.6)**
- PhosStop (4906845001, Roche, Basil, Switzerland) (*see* **Note 4.1**)
- QIAamp DNA Mini Kit (51304; Qiagen, Hilden, Germany)
- Ethanol Absolute (3406.500; J.T.Baker, Phillipsburg, NJ, USA) (*see* **Notes 4.1 and 4.7**)
- QIAamp DNA Blood Mini Kit (51104; Qiagen)
- PowerUp™ SYBR™ Green Master Mix for qPCR (A25780; Applied Biosystems™, Waltham, MA, USA)
- Primers (Table 1) (*see* **Notes 4.8**)
- *Staphylococcus aureus* (43300, ATCC collection)
- Brain Heart Infusion BHI broth (620008; Liofilchem)
- Columbia CNA agar plates (11024; Liofilchem)

2.3 Equipment

- Burker chamber (0.100 mm depth, 0.0025 mm^2) (Fortuna; Marienfeld, Germany)
- Microcentrifuge (5418 R; Eppendorf, Hamburg, Germany)
- Instrument for real-time PCR (Mastercycle Realplex^2; Eppendorf)
- Laminar Flow Hood with HEPA filter for cell culture and bacteria
- Incubator with CO_2 for cell culture (BB15; Thermo Scientific)
- Inverted microscope for cell observation (Eclipse TS100; Nikon, Amstelveen, Netherland)

Table 1 Sequences and annealing temperatures for the used primers.

Gene	Primer	Sequence	Annealing temperature
mtDLoop	Fw	5'-GTCCCTTGACCACCATCCTC	60 °C
	Rv	5'-GTAGCACTCTTGTGCGGGAT	60 °C
mtND4	Fw	5'-CCCTCGTAGTAACAGCCATTCTC	60 °C
	Rv	5'-GCACTGTGAGTGCGTTCGTAGT	60 °C
GAPDH	Fw	5'-AGCCACATCGCTCAGACAC	60 °C
	Rv	5'-GCCCAATACGACCAAATCC	60 °C

- Thermo-shaker with Cooling for Microtubes and Microplates (Temperature setting range: +4° C to +100° C) (PCMT; Grant-bio, Royston, UK)
- Incubator for bacteria (BB15; Thermofisher Scientific)
- UV spectrophotometer (Spectronic 20 Genesys; Thermofisher Scientific)
- GraphPad Prism software (version 10.1.1; GraphPad) (*see* **Note 4.9**)

3. Methods

3.1 Reagent preparation

- Trypsin-EDTA 1×: Trypsin-EDTA solution 10× was diluted 1:10 with 1× D-PBS. (*see* **Note 4.3**)
- cmtDNA extraction solution: cmtDNA extraction solution was prepared with 220 mM mannitol, 70 mM sucrose, 20 mM HEPES-KOH pH 7.5, 1 mM EDTA, 2 mg/mL BSA, 1:100 protease inhibitors, and 1:10 PhosStop in distilled water. Add 2 μg/mL BSA before the use.
- Primer reconstitution: Lyophilized primers were reconstituted with MilliQ water to final concentration 10 μM.
- BHI broth: Suspend 37 g of the powder in 1 L of deionized water and heat to boil shaking frequently until completely dissolved. Sterilize before use.

3.2 Cell culture and treatments

- HaCaT cells were cultured in complete culture medium, consisting of HG-DMEM supplemented with 10% FBS, and 1% penicillin/streptomycin, and maintained at 37 °C, 5% CO_2 in the incubator. (*see* **Note 4.3**)
- One day before the *S. aureus* infection, cells were detached by removing the culture medium, washed twice in PBS, and incubated in pre-warmed 1× Trypsin-EDTA solution at 37 °C for 10 min. Suspended cells in Trypsin-EDTA solution were transferred into a 15 mL conical tube and trypsin was inactivated by adding culture medium. (*see* **Note 4.3**)
- 10 μL cell suspension diluted 1:1 in Trypan Blue were inserted into a Burker chamber and live cells (the not-blue ones) were counted with an inverted microscope.
- HaCaT were seeded into 6-well plates at cell density 5×10^5 cell/cm^2 and incubated up to 24 h before infection. Two wells (a total of 1×10^6 cells) were considered for each condition.

- One day before infection, *S. aureus* ATCC 43300 was seeded into CNA agar plates and incubated O/N at 37 °C. (*see* **Note 4.10**)
- One colony of *S. aureus* was subcultured in a 15 mL tube containing 3 mL of BHI broth and incubated for 3 h at 37 °C to reach exponential growth phase.
- Broth culture was diluted in HG-DMEM supplemented with 10% FBS (w/o 1% penicillin/streptomycin) to reach an $O.D._{600}$ of 0.1 resulting in a concentration of $\sim 1 \times 10^8$ CFU/mL.
- Bacterial culture was further diluted 1:2.5 in DMEM to achieve a final concentration of 4×10^7 CFU/mL.
- Cell media of HaCaT cells was removed and 500 μL of bacterial culture was added to each well (final MOI 20×). Cells were incubated at 37 °C, 5% CO_2 for 3 h in the incubator. (*see* **Note 4.11)**
- 500 μL of HG-DMEM supplemented with 10% FBS and 2% penicillin/streptomycin were added to each well to stop bacterial growth.
- Simultaneously to the infection, cells for positive control (CTR+) were stimulated with 20 μg/mL LPS (CTR+).
- Cells were collected after 2 (t2), 4 (t4), and 6 (t6) hours upon the medium change after the MRSA infection, setting three different time points. Cells without stimuli were used as control (CTR) and collected at t6, as cells for CTR+. For the collection, cells were washed with PBS 1×, detached with a cell scraper in 0.75 mL PBS 1× per well, and the cells considered for the same condition were collected together into a 1.5 mL microcentrifuge tube. Centrifuge at 1300 rpm for 5 min at 4 °C was performed to remove the supernatant, before resuspending pellets in 1 mL PBS 1×. 200 μL of cell suspension were stored for whole cell (WC) mtDNA extraction, whilst the remaining 800 μL were saved up for cytosolic mtDNA (cmtDNA).

3.3 mtDNA extraction from whole cells (see Notes 4.2)

WC mtDNA extraction was performed using the QIAamp DNA Mini Kit, mainly according to the manufacturer's instructions with some modifications.

- Cells were pelleted at 1300 g for 5 min at 4 °C and resuspended in 100 μL Buffer ATL and 10 μL Proteinase K (included in the kit). (*see* **Note 4.12**) The suspension was vortexed and incubated at 56 °C for 10 min.
- 100 μL Buffer AL were added, before vortexing for 15 s.
- 50 μL 100° Ethanol were added, before vortexing for 15 s and incubating for 3 min at room temperature (RT) before centrifuging to remove drops from the lid. (*see* **Note 4.7**)

- The liquid was pipetted onto a spin column inserted into a 2 mL collection tube, centrifuged at 6000 *g* for 1 min, and the flow-through was discarded.
- The first wash was performed by adding 500 µL Buffer AW1 onto the spin column and placing it into a new 2 mL collection tube, centrifuging at 6000 *g* for 1 min, before discarding the flow-through.
- 500 µL Buffer AW2 were placed onto the spin column with a new 2 mL collection tube, centrifuged at 20,000 *g* for 3 min, and the flow-through was discarded again, to complete the second wash.
- To collect DNA, 20 µL Buffer AE were pipetted onto the spin column (inserted into a 1.5 mL collection tube) and incubated for 1 min, before centrifuging at 6000 *g* for 1 min. (*see* **Note 4.13**)

3.4 mtDNA extraction from the cytosol (see Note 4.14)

- To extract the cytosolic mtDNA, the remaining 800 µL suspension was centrifuged at 1300 *g* and the obtained cell pellet was resuspended in 250 µL cmtDNA Extraction Solution.
- Homogenization was performed with a 26 G needle syringe 25 times. A first centrifugation was performed at 1000 *g* for 15 min to remove any remaining whole cell, and the supernatant was collected.
- A second centrifugation was executed at 10,000 *g* for 10 min to collect the supernatant containing the cytosolic part. (*see* **Note 4.15**).
- The obtained supernatant was processed (200 µL) with QIAamp DNA Blood Mini Kit. (*see* **Note 4.2**).
  - 20 µL of Proteinase K (included in the kit) were added to the supernatant, and mixed, before adding 200 µL of Buffer AL and mixing thoroughly to obtain a homogeneous solution. (*see* **Note 4.11**)
  - The sample was incubated at 56 °C for 10 min.
  - 200 µL Ethanol 100° were added to the sample and vortexed for 15 s. A brief centrifuge was performed to remove drops from the lid. (*see* **Note 4.7**)
  - The mixture was added to a spin column (in a 2 mL collection tube) and centrifuged at 6000 *g* for 1 min. The flow-through and the collection tube were discarded.
  - The spin column was provided of a new 2 mL collection tube and washed with 500 µL Buffer AW1. The sample was centrifuged at 6000 *g* for 1 min, and the flow-through and the collection tube were discarded.

- A new wash was performed adding 500 µL Buffer AW2 and centrifuging at 20,000 *g* for 3 min. The flow-through and the collection tube were discarded, and an additional centrifuge at 20,000 *g* for 1 min was performed to remove any buffer remaining. After discarding the flow-through and the collection tube, the spin column was placed into a 1.5 mL collection tube.
- 200 µL Buffer AE were added to the spin column and incubated for 1 min, before centrifuging at 6000 g for 1 min. (*see* **Note 4.13**)

3.5 RT-PCR (see Notes 4.2 and 4.16)

RT-PCR was performed according to the PowerUp™ SYBR™ Green Master Mix for qPCR's manual. DNA from whole cells was used to evaluate the expression of mtND4 and mtDLoop, the two target genes, and GAPDH, as the housekeeping gene. cmtDNA extracts were used to evaluate the mtND4 and mtDLoop expression in the cytosol. Fig. 1 contains a schematic representation of the final template for each gene.

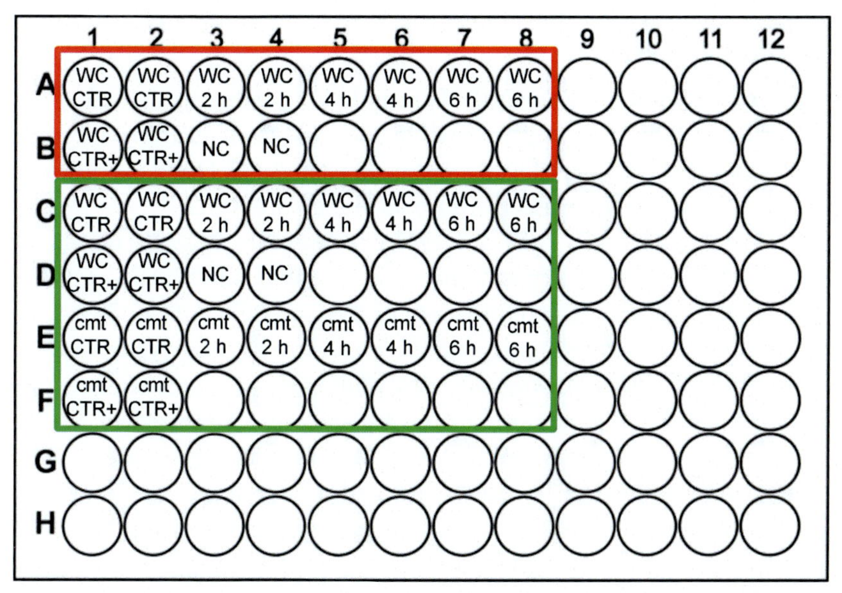

Fig. 1 Schematic representation of the final template for each gene. In the red rectangle, the samples analyzed for GAPDH; in the green rectangle, the samples analyzed for mtND4 or mtDLoop.

- For each analyzed gene, a complete mix constituted of Master Mix, forward (Fw) and reverse (Rv) primers, and MilliQ water was prepared. The mix was prepared considering for each sample with 5 μL Master Mix, 0.5 μL Fw primer 10 μM (final concentration 500 nM), 0.5 μL Rv primer 10 μM (final concentration 500 nM), and 2 μL MilliQ water. (*see* **Notes 4.17 and 4.18**)
- 2 μL DNA were loaded into each tube of MicroAmp™ Optical 8-Cap Strips. Two tubes for negative controls were loaded with 2 μL MilliQ water for each analyzed gene. (*see* **Note 4.18**)
- 8 μL complete mix were added in each tube pipetting to properly amalgam the mix and the sample. The total volume in each tube was 10 μL.
- The tubes were carefully closed, and the strips were spun to deposit the samples at the bottom and remove any bubbles.
- The strips were located into the thermocycler machine, and the reaction parameters were set up. The reaction was performed at 50 °C for 2 min to activate UGD and at 90 °C for 2 min to activate DNA polymerase. After, 40 cycles (95 °C for 15 s and 60 °C for 1 min) were performed to denature DNA and annealing/extending.

3.6 Data analysis

- For each sample, a mean Cycle Threshold (Ct) of the two replicates was calculated. The first step to analyze the cmtDNA relative expression was to normalize the mtDNA expression from the WC and the obtained mtDNA amount was used to normalize the cmtDNA.

  More specifically, the cmtDNA expression was calculated by the $2^{-\Delta\Delta Cq}$ method, considering: Cq (Ct(WC mtDNA) – Ct(WC GAPDH); ΔCq (Ct(cmtDNA)—Cq); ΔΔCq (ΔCq - ΔCq CTR).
- The results of three independent experiments were examined in GraphPad Prism v.10 software. A Column file was created, and two independent data tables (one for each gene) were generated.
- The relative expressions for CTR, 2 h, 4 h, 6 h, and CTR + were inserted into the correspondent data table and analyzed by the One-way ANOVA method. "No matching or pairing" and ordinary ANOVA test options were selected in the "Experimental Design" section. The mean of each column was compared with the mean of the CTR column ("Multiple Comparisons" section).
- The results were represented as the mean ± standard deviation (SD) (Fig. 2).

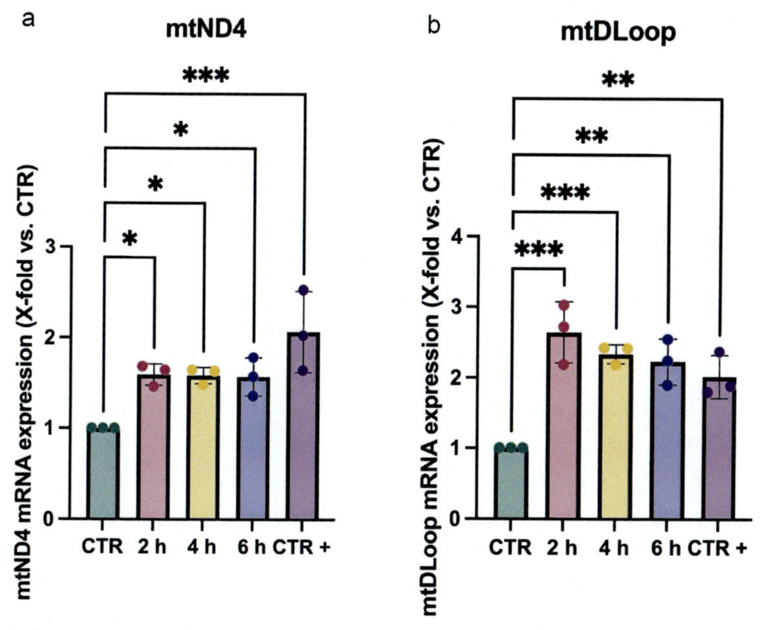

Fig. 2 Representative histograms for (A) mtND4 and (B) mtDLoop relative expression. (mtND4 One-way ANOVA $P = 0.0035$, mtDLoop One-way ANOVA $P = 0.0004$; * $P \leq 0.05$, ** $P \leq 0.01$, *** $P \leq 0.001$).

4. Notes

4.1. Catalog number and provider for disposable and reagents are indicated as a reference, but an equivalent product can be purchased from a variety of sources at a similar cost.

4.2. Nuclease-free plasticware, reagents, and filtered tips are highly recommended to avoid DNAse contamination.

4.3. DMEM, FBS, Trypsin-EDTA, and cells are all considered as non-hazardous (NONH) but should be nonetheless manipulated with appropriate PPE.

4.4. Potassium hydroxide is corrosive and irritant (H290) and causes severe skin and eye burns (H314). Do not breathe in dust/fumes/gases/mist/vapors/aerosols and wear protective gloves and eye/face protection. In case of contact, rinse thoroughly with water for several minutes.

4.5. EDTA causes skin corrosion/irritation (H315) and severe eye injury/eye irritation (H319) and may irritate the respiratory tract

(H335). Avoid breathing dust/fumes/gases/mist/vapors/aerosols. In case of contact: rinse thoroughly with water.

4.6. Protease inhibitors are corrosive and irritant and cause severe skin and eye burns (H314) and can irritate the respiratory tract (H335). Do not breathe in dust/fumes/gases/mist/vapors/aerosols and wear protective gloves and eye/face protection. In case of contact: rinse skin/eye thoroughly with water.

4.7. Ethanol is highly flammable in liquid and vapor (H225) and causes serious eye irritation (H319), and hence it should be manipulated by wearing appropriate recommended PPE and at a distance from open flames, as well as potential sources of heat and sparkles, and it should be stored in dedicated cabinets for flammables.

4.8. mtDNA detection could be also assessed by different primers for the mentioned genes and/or other mitochondrial genes.

4.9. A different software version could be used; however, some features and functions could be different.

4.10. *S. aureus* culture can also be performed in liquid media such as BHI and subculture can be made by diluting 1:100 O/N broth in fresh BHI.

4.11. Alternative to MRSA infection, cells could be treated with MRSA supernatant (MRSA-S), obtained inoculating *S. aureus* in 10 mL BHI broth o.n. at 37 °C, then centrifuging at 8000*g* for 10 min, and filtering the obtained supernatant with a 0.22 μm syringe filter. For cell treatment, MRSA-S can be diluted 1:5 in complete HG-DMEM.

4.12. Proteinase K could lead to respiratory sensitization and danger in case of suction (H334). Manipulate under a chemical hood.

4.13. Obtained DNA can be processed immediately. Alternatively, it can be stored at −20 °C for short-term storage or at −80 °C for long-term storage.

4.14. During the cmtDNA extraction, the suspension was maintained in ice during the process and the centrifuge was set at 4 °C.

4.15. Purity of cytosolic extract could be assessed by Western Blotting, evaluating the expression of mitochondrial markers (*i.e.*, TOM20, TIM23, and HSP60).

4.16. The primers and DNA should be thawed in ice before use.

4.17. Because of possible pipetting errors, the complete mix should be prepared considering an excess (*e.g.*, calculating one more sample) and considering negative controls.

4.18. Each sample should be double-loaded, to avoid false results due to possible pipetting errors.

5. Concluding remarks

The major advantages of employing RT-PCR for assessing cytosolic mtDNA levels consist in the relative low cost of both reagents and technical platforms, as well as in the high processivity. Key points for the success of the experiment are the collection of pure cytosolic fractions (without mitochondrial contamination), as well as proper measuring of total cellular mtDNA in the whole cell lysate as standardization. In general, results obtained using PCR-based methods should be accompanied by analysis through immunofluorescence microscopy (Bryant, Lei, VanPortfliet, Winters, & West, 2022; Licini et al., 2024; Newman et al., 2024; Sato et al., 2021), using anti-DNA antibodies or specific fluorescent probes (*i.e.*, PicoGreen dye) for marking DNA, followed by calculation of the non-mitochondrial fluorescent dots (cytosolic mtDNA nucleoids). Therefore, combining the method described here with a confocal-based approach represents, at today, the most reliable strategy for assessing the extent of mtDNA release in human cells.

Acknowledgments

S.M. is supported by Progetti di Rilevante Interesse Nazionale (PRIN2022ALJN73 and PRINP202243ZAR), and local funds from Marche Polytechnic University (Ancona, Italy). The research leading to these results has received funding from the European Union - NextGenerationEU through the Italian Ministry of University and Research under PNRR—M4C2-I1.3 Project PE_00000019 "HEAL ITALIA" to Saverio Marchi (S.M.), CUP I33C22006900006 (Marche Polytechnic University).

Conflicts of interest

The authors declare that they have no competing interests.

References

Akbal, A., Dernst, A., Lovotti, M., Mangan, M. S. J., McManus, R. M., & Latz, E. (2022). How location and cellular signaling combine to activate the NLRP3 inflammasome. *Cellular & Molecular Immunology*, *19*(11), 1201–1214. https://doi.org/10.1038/s41423-022-00922-w.

Bryant, J. D., Lei, Y., VanPortfliet, J. J., Winters, A. D., & West, A. P. (2022). Assessing mitochondrial DNA release into the cytosol and subsequent activation of innate immune-related pathways in mammalian cells. *Current Protocols*, *2*(2), e372. https://doi.org/10.1002/cpz1.372.

Carlos, D., Costa, F. R. C., Pereira, C. A., Rocha, F. A., Yaochite, J. N. U., Oliveira, G. G., et al. (2017). Mitochondrial DNA activates the NLRP3 Inflammasome and predisposes to type 1 diabetes in murine model. *Frontiers in Immunology*, *8*. https://doi.org/10.3389/fimmu.2017.00164.

Chen, H., Zhang, J., He, Y., Lv, Z., Liang, Z., Chen, J., et al. (2022). Exploring the role of Staphylococcus aureus in inflammatory diseases. *Toxins*, *14*(7), 464. https://doi.org/10.3390/toxins14070464.

Cosentino, K., Hertlein, V., Jenner, A., Dellmann, T., Gojkovic, M., Peña-Blanco, A., et al. (2022). The interplay between BAX and BAK tunes apoptotic pore growth to control mitochondrial-DNA-mediated inflammation. *Molecular Cell*, *82*(5), 933–949.e9. https://doi.org/10.1016/j.molcel.2022.01.008.

Guilbaud, E., & Galluzzi, L. (2022). Adaptation to MOMP drives cancer persistence. *Cell Research*, *33*(2), 93–94. https://doi.org/10.1038/s41422-022-00729-4.

Howden, B. P., Giulieri, S. G., Wong Fok Lung, T., Baines, S. L., Sharkey, L. K., Lee, J. Y. H., et al. (2023). Staphylococcus aureus host interactions and adaptation. *Nature Reviews Microbiology*, *21*(6), 380–395. https://doi.org/10.1038/s41579-023-00852-y.

Kim, J., Kim, H.-S., & Chung, J. H. (2023). Molecular mechanisms of mitochondrial DNA release and activation of the cGAS-STING pathway. *Experimental & Molecular Medicine*, *55*(3), 510–519. https://doi.org/10.1038/s12276-023-00965-7.

Licini, C., Morroni, G., Lucarini, G., Vitto, V. A. M., Orlando, F., Missiroli, S., et al. (2024). ER-mitochondria association negatively affects wound healing by regulating NLRP3 activation. *Cell Death & Disease*, *15*(6), 407. https://doi.org/10.1038/s41419-024-06765-9.

Marchi, S., Guilbaud, E., Tait, S. W. G., Yamazaki, T., & Galluzzi, L. (2023). Mitochondrial control of inflammation. *Nature Reviews Immunology*, *23*(3), 159–173. https://doi.org/10.1038/s41577-022-00760-x.

Marchi, S., Morroni, G., Pinton, P., & Galluzzi, L. (2022). Control of host mitochondria by bacterial pathogens. *Trends in Microbiology*, *30*(5), 452–465. https://doi.org/10.1016/j.tim.2021.09.010.

McArthur, K., Whitehead, L. W., Heddleston, J. M., Li, L., Padman, B. S., Oorschot, V., et al. (2018). BAK/BAX macropores facilitate mitochondrial herniation and mtDNA efflux during apoptosis. *Science*, *359*(6378), eaao6047. https://doi.org/10.1126/science.aao6047.

Missiakas, D., & Winstel, V. (2021). Selective host cell death by Staphylococcus aureus: A strategy for bacterial persistence. *Frontiers in Immunology*, *11*, 621733. https://doi.org/10.3389/fimmu.2020.621733.

Newman, L. E., Weiser Novak, S., Rojas, G. R., Tadepalle, N., Schiavon, C. R., Grotjahn, D. A., et al. (2024). Mitochondrial DNA replication stress triggers a pro-inflammatory endosomal pathway of nucleoid disposal. *Nature Cell Biology*, *26*(2), 194–206. https://doi.org/10.1038/s41556-023-01343-1.

Pastar, I., Stojadinovic, O., Yin, N. C., Ramirez, H., Nusbaum, A. G., Sawaya, A., et al. (2014). Epithelialization in wound healing: A comprehensive review. *Advances in Wound Care*, *3*(7), 445–464. https://doi.org/10.1089/wound.2013.0473.

Riley, J. S., Quarato, G., Cloix, C., Lopez, J., O'Prey, J., Pearson, M., et al. (2018). Mitochondrial inner membrane permeabilisation enables mt DNA release during apoptosis. *The EMBO Journal*, *37*(17), e99238. https://doi.org/10.15252/embj.201899238.

Riley, J. S., & Tait, S. W. (2020). Mitochondrial DNA in inflammation and immunity. *EMBO Reports*, *21*(4), e49799. https://doi.org/10.15252/embr.201949799.

Sato, A., Buque, A., Yamazaki, T., Bloy, N., Petroni, G., & Galluzzi, L. (2021). Immunofluorescence microscopy-based assessment of cytosolic DNA accumulation in mammalian cells. *STAR Protocols*, *2*(2), 100488. https://doi.org/10.1016/j.xpro.2021.100488.

Soe, Y. M., Bedoui, S., Stinear, T. P., & Hachani, A. (2021). Intracellular *Staphylococcus aureus* and host cell death pathways. *Cellular Microbiology*, *23*(5). https://doi.org/10.1111/cmi.13317.

Vringer, E., Heilig, R., Riley, J. S., Black, A., Cloix, C., Skalka, G., et al. (2024). Mitochondrial outer membrane integrity regulates a ubiquitin-dependent and NF-κB-mediated inflammatory response. *The EMBO Journal*, *43*(6), 904–930. https://doi.org/10.1038/s44318-024-00044-1.

Wang, X., Li, H., Wang, J., Xu, H., Xue, K., Liu, X., et al. (2023). Staphylococcus aureus extracellular vesicles induce apoptosis and restrain mitophagy-mediated degradation of damaged mitochondria. *Microbiological Research*, *273*, 127421. https://doi.org/10.1016/j.micres.2023.127421.

Worster, B., Zawora, M. Q., & Hsieh, C. (2015). Common questions about wound care. *American Family Physician*, *91*(2), 86–92.

Zhang, X., Hu, X., & Rao, X. (2017). Apoptosis induced by Staphylococcus aureus toxins. *Microbiological Research*, *205*, 19–24. https://doi.org/10.1016/j.micres.2017.08.006.

Zhang, W., Li, G., Luo, R., Lei, J., Song, Y., Wang, B., et al. (2022). Cytosolic escape of mitochondrial DNA triggers cGAS-STING-NLRP3 axis-dependent nucleus pulposus cell pyroptosis. *Experimental & Molecular Medicine*, *54*(2), 129–142. https://doi.org/10.1038/s12276-022-00729-9.

CHAPTER SEVEN

Monitoring cellular dynamics upon infection using a holotomography-based approach

Ilaria Nunzi[a,†], Gloria D'Achille[b,†], Nada Dhaouadi[a], Fabio Marcheggiani[a], Caterina Licini[a], Mariangela Di Vincenzo[a], Monia Orciani[a], Gianluca Morroni[b,*], and Saverio Marchi[a,c,*]

[a]Department of Clinical and Molecular Sciences, Marche Polytechnic University, Ancona, Italy
[b]Microbiology Unit, Department of Biomedical Sciences and Public Health, Marche Polytechnic University, Ancona, Italy
[c]Advanced Technology Center for Aging Research, IRCCS INRCA, Ancona, Italy
*Corresponding authors: e-mail address: g.morroni@univpm.it; s.marchi@staff.univpm.it

Contents

Abstract

Many intracellular bacteria interfere with mitochondrial dynamics or target other organelles, thereby inducing a specific cellular response that could emerge as a strategy of the pathogen to ensure its survival, or as a form of defense employed by the host cell to restrict dissemination. In this context, the concomitant monitoring of both pathogen migration and (intra)cellular dynamics in live cells emerges as a pivotal aspect for the comprehension of the infection sequence and to visualize the pathogen-mediated remodeling that could occur to the entire cellular system. Holotomographic microscopy can

† Co-first author.

Methods in Cell Biology, Volume 194
ISSN 0091-679X
https://doi.org/10.1016/bs.mcb.2024.12.003

be used to achieve this goal, allowing the simultaneous analysis of both bacterial movement and intracellular alteration for extended periods of time, with high spatial resolution and avoiding side-effects due to phototoxicity.

Here we provide a holotomography-based approach to detect *Listeria monocytogenes* dynamics and its effects on the entire cellular system at morphological level.

1. Introduction

Listeria monocytogenes is a ubiquitous Gram-positive rod that can survive in different environments, adapting to various environmental conditions including low temperatures, high salt concentrations, and acidity (Koopmans, Brouwer, Vazquez-Boland, & van de Beek, 2023). *L. monocytogenes* can infect people and cause listeriosis, an illness ranging from subclinical disease to invasive and life-threatening forms. Listeriosis is a food-borne infection as *L. monocytogenes* can contaminate various foods such as meat, dairy products, vegetables, fruit, and prepared food (Schlech, 2019), making this species one of the major food-borne pathogens responsible for numerous outbreaks during recent years. In the most severe infections, invasive *Listeria* infections lead to sepsis, central nervous system (CNS) infections, and maternal-fetal infections (Lecuit, 2020). In all these cases, the mortality rates are quite high: systemic listeriosis has a mortality rate ranging from 21% to 46%, while adult meningoencephalitis mortality reached 50% (Charlier et al., 2017).

Several virulence factors contribute to *Listeria* pathogenesis, including toxins (listeriolysin-O, phospholipases) (Hamon, Ribet, Stavru, & Cossart, 2012), actin remodeling protein (ActA), surface proteins for adhesion (Kocks et al., 1992; Tilney & Portnoy, 1989), and invasins (InlA and InlB) for internalization in eukaryotic cells (Lecuit et al., 2001). Indeed, the pathogenesis of *L. monocytogenes* infections involves the translocation through the intestinal barrier, invading M-cells or the enterocytes in a InlA/InlB-mediated process and causing gastroenteritis (Koopmans et al., 2023). During this phase, *Listeria* InlA binds the E-cadherin receptors expressed by the epithelial cells and promotes the internalization of the pathogen. Once inside the enterocytes, *L. monocytogenes* is able to escape from the internalization vacuole and reaches the cytoplasm, starting a cytosolic replication. Similarly, *Listeria* is also phagocytized by professional phagocytic cells but, through its toxins, the bacteria can escape the phagosome, promoting its survival, replication, and diffusion into the host (Ruan, Rezelj, Bedina Zavec, Anderluh, & Scheuring, 2016).

This passage is essential for the dissemination of the pathogen in the mesenteric lymph nodes and then in the bloodstream. In this context, *Listeria* acquires different mechanisms to enable immune escape, similarly to what described for neoplastic cells (Galassi, Chan, Vitale, & Galluzzi, 2024), leading to bacterial persistence and neuroinvasion (Ling et al., 2021; Maudet et al., 2022).

Other stratagems developed by *Listeria* to elude host cell recognition include the inhibition of autophagy (Tattoli et al., 2013; Yang et al., 2024) and perturbation of organelle functions (Lebreton, Stavru, & Cossart, 2015). Indeed, one of the primary intracellular targets of *Listeria* activity is represented by the mitochondrial compartment (Marchi, Morroni, Pinton, & Galluzzi, 2022). *L. monocytogenes* infection induces a rapid and transient fragmentation of the mitochondrial network, with concomitant mitochondrial depolarization and reactive oxygen species (ROS) production (Carvalho et al., 2020; Stavru, Bouillaud, Sartori, Ricquier, & Cossart, 2011; Stavru, Palmer, Wang, Youle, & Cossart, 2013). Importantly, *L. monocytogenes* infection is restricted by alteration of key mitochondrial functions, including abolition of mitochondrial Ca^{2+} entry via the mitochondrial calcium uniporter (MCU) (Li et al., 2021), inhibition of mitophagy (Galluzzi et al., 2008; Zhang et al., 2019), increased mitochondrial respiration (Spier et al., 2021), and increased mitochondrial fission through upregulation of MIC10, a core subunit of the mitochondrial contact site and cristae organizing system (MICOS) complex (Carvalho et al., 2020). Thus, monitoring mitochondrial dynamics during *L. monocytogenes* infection is a crucial aspect to understand the pathogenic mechanisms of *Listeria*, or to decipher how the cell promptly reacts to bacterial invasion.

Holotomographic microscopy (HTM) produces high-content refractive index (RI) images that reveal sophisticated biological processes and multiple intracellular elements at high spatial resolution and ultralow-power light source, without the employment of fluorescent indicators (Sandoz, Tremblay, van der Goot, & Frechin, 2019). The HTM-based capturing of label-free timelapse images of cells and their organelles allows to give insights into phenotypic changes that occur upon different stimuli, including bacterial infection.

Here we report a method to monitor the intracellular pathogen *L. monocytogenes* in HCT116 cells using a HTM-based approach, to visualize the localization of bacteria, as well as the cellular response and morphological changes.

2. Materials

2.1 Common disposables (see Note 4.1)

- 10, 200 and 1000 μL pipette tips (S1111-3000, S1111-0006, S1111-6000; Starlab S.r.l., Milan, Italy).
- Flask T-75 CytoOne®, TC treated, vented (CC7682-4875; Starlab S.r. l., Milan, Italy) (***see Note 4.1 and 4.2***).
- 1, 2, 5, 10, 25 mL serological pipettes (E4860-0001, E4860-0002, E4860-0005, E4860-0010; Starlab S.r.l., Milan, Italy).
- 1.5 mL microcentrifuge tubes (S1615-5550; Starlab S.r.l., Milan, Italy).
- 15 mL centrifuge tubes (E1415-0100; Starlab S.r.l., Milan, Italy).
- μ-Dish 35 mm, 1.5 polymer coverslip, tissue culture-treated, sterilized (81156; ibidi GmbH, Gräfelfing, Germany).
- Cuvette (0193800; Kartell™)
- Falcon® 15 mL Conical Bottom Polypropylene Centrifuge Tubes, with flat cap, sterile (525-0604; Avantor)
- Metal loop
- Sterile swabs (710-0186; Avantor)

2.2 Cells and reagents (see Note 4.1)

- HCT116 cells (RRID:CVCL_0291; Cellosaurus database)
- DMEM High Glucose (4.5 g/L), w/o L-Glutamine, with Sodium Pyruvate (SIAL-DMEM-HPXA; Sial, Rome, Italy) (***see Note 4.3***)
- Fetal Bovine Serum (FBS) South America Superior (YOURSIAL-FBS-SA; Sial, Rome, Italy) (***see Note 4.3***)
- Penicillin/Streptomycin 100× (ECB3001D; Euroclone, Pero, Italy) (***see Note 4.4***)
- Corning™ L-Glutamine solution cellgro™ (15393631; Corning) (***see Note 4.3***)
- Dulbecco's Phosphate-Buffered Saline (D-PBS), 1× without calcium and magnesium (21-031-CV; Corning)
- Trypsin-EDTA solution 10× (T4174; Sigma-Aldrich, St. Louis, MO, USA) (***see Note 4.3***)
- Trypan Blue stain 0.4% (T10822; Invitrogen™, Waltham, MA, USA)
- Brain Heart Infusion BHI Broth (610008; Liofilchem) (***see Note 4.5***)

- Columbia CNA agar plates (11024; Liofilchem) (***see Note 4.6***)
- Gentamicin (G1264; Sigma)
- *L. monocytogenes* (13932: ATCC collection) (***see Note 4.7***)

2.3 Equipment

- Bunsen burner
- Incubator for bacteria (BB15; Thermo Fisher Scientific)
- UV/Visible Spectrophotometer (DU 530 Life Science; Beckman)
- Bürker chamber (0.100 mm depth, 0.0025 mm^2) (Fortuna; Marienfeld, Germany) (***see Note 4.1***)
- Laminar Flow Hood with HEPA filter for cell culture and bacteria (***see Note 4.1***)
- Incubator with CO_2 for cell culture (BB15; Thermo Scientific) (***see Note 4.1***)
- Inverted microscope for cell observation (Eclipse TS100; Nikon, Amstelveen, Netherland) (***see Note 4.1***)
- 3D Cell Explorer-fluo holotomographic microscope (Nanolive SA; Ecublens, Switzerland)
- UNO-T-H-CO_2 top-stage incubator with CO_2 controller (OkoLab S.R.L., Pozzuoli, ITA)
- STEVE full software (version 1.6.3496, Nanolive SA)
- EVE analytics software (version 2.1.0.1987, Nanolive SA)

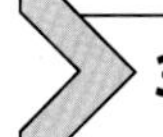

3. Methods

3.1 Bacterial growth and inoculum preparation

- The day before infection, *L. monocytogenes* ATCC 13932 was seeded in CNA plates and incubated overnight at 37 °C.
- One colony of *L. monocytogenes* was placed in a 15 mL tube containing 3 mL of BHI broth and incubated at 37 °C for at least 3 h to reach exponential growth.
- 1 mL of BHI broth was used to perform the blank at the spectrophotometer setting λ at 650 nm.
- The optical density (OD) of 1 mL of *L. monocytogenes* broth culture was measured and diluted in DMEM to reach an OD650 of 0.1 corresponding to a bacterial concentration of $\sim 1 \times 10^8$ CFU/mL.
- Bacterial culture was further diluted in DMEM to achieve a final concentration of 3×10^6 CFU/mL.

- 100 μL of bacterial culture was prepared to be added to the cells, obtaining a multiplicity of infection (MOI) of 20.

3.2 Cell analysis upon infection using holotomographic approach

3.2.1 *Reagent preparation*

- Trypsin-EDTA solution 10× was diluted 1:10 with 1× D-PBS to obtain Trypsin-EDTA solution 1× (***see Note 4.3***).

3.2.2 *Cell culture and imaging*

- HCT116 cells are maintained at 37 °C under 5% CO_2 in the incubator in DMEM HG supplemented with 10% FBS, 1% L-Glutamine 200 mM and 1% penicillin/streptomycin 100× (complete culture medium) (***see Note 4.3***).
- One day before the experiment cells are detached by removing exhausted culture medium, are washed twice with D-PBS 1× and incubated in pre-warmed 1× Trypsin-EDTA solution at 37 °C for 5 min.
- Trypsin is inactivated with complete culture medium and detached cells are collected in a 15 mL centrifuge tube.
- Cells are counted in an inverted microscope through the use of Bürker chamber, where is inserted 10 μL cell suspension diluted 1:1 in Trypan blue.
- 1.5×10^4 cells are seeded in μ-Dish 35 mm with a growth area of 3.5 cm^2 and incubated for 24 h.
- The day of the experiment the complete culture medium in the dish is substituted with 900 μL of fresh DMEM HG + L-Glutamine (no FBS and Penicillin/Streptomycin).
- The dish is placed in the top-stage incubator, and allowed to equilibrate for 30 min.
- 100 μL of bacterial culture (MOI 20) is added drop by drop to the cells. The images in high resolution are collected every 1 min for 1 h and monitored with the use of STEVE software (***see Note 4.8***).
- After 1 h of infection, 1 mL of FBS-containing DMEM supplemented with 40 μg/mL gentamicin is added to kill extracellular bacteria. The images are collected for other 5 h (total: 6 h).

3.2.3 *Data processing*

- After 6 h, the images have been processed using the EVE analytics software, which offers a segmentation and analysis solution specific to Nanolive's content (Fig. 1).

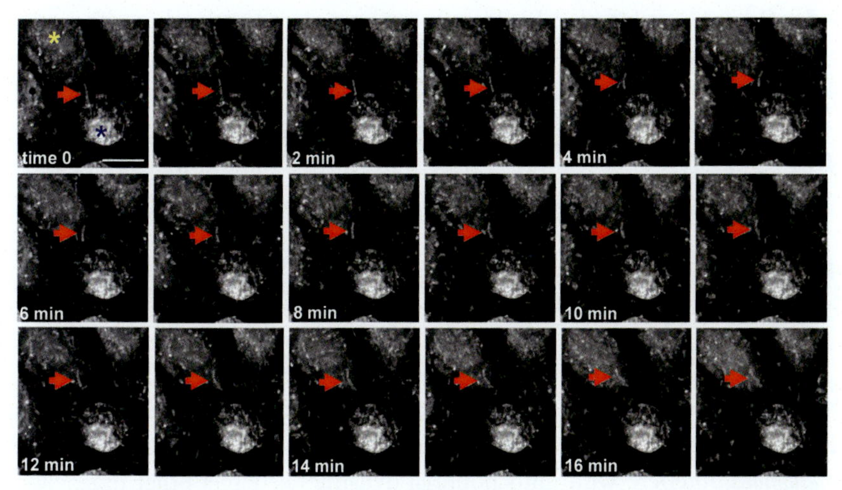

Fig. 1 Visualization of living *Listeria monocytogenes* using a 3D tomographic approach. Visualization of unstained *Listeria monocytogenes* over time. Red arrows showed a single Listeria coming out of a died cell (purple asterisk) and entering into an adjacent cell (yellow asterisk). Scale bar: 10 μm.

Alternatively, holographic reconstruction processing could be performed using the Fiji ImageJ software (Schindelin et al., 2012). In brief, an export can be carried out within the software STEVE, in order to transform RI volumes into .tiff format. In this manner, the files are readable by the Fiji software and the 3D RI volumes in .tiff format could be processed in batch for specific purposes. Finally, the 3D RI images were converted into 2D RI maps using maximum intensity projections and saved as .tiff files.

4. Notes

4.1. Equivalent products can be used by alternative providers at similar cost. The catalogue number and reagents are indicated as a reference.

4.2. The size of the flasks can be smaller (25 cm^2) or bigger (175 cm^2) according to the amount of cells needed.

4.3. DMEM, FBS, L-Glutamine solution and Trypsin-EDTA are considered as not hazardous (NONH) but it is recommended to manipulate them with appropriate PPE.

4.4. Penicillin/Streptomycin should be manipulated with appropriate PPE to avoid allergic skin reaction (H317).

4.5. Brain Heart Infusion BHI broth can be replaced with other rich media like Tryptic Soy Broth.
4.6. Columbia CNA agar plates can be replaced with other agarose-based media like Brain Heart Infusion Agar or Tryptic Soy Agar.
4.7. *L. monocytogenes* culture can be performed in liquid media like BHI broth and subcultured diluting 1:100 the overnight bacterial culture in fresh BHI broth.
4.8. The time of acquisition can be changed.

5. Concluding remarks

In addition to *L. monocytogenes*, many other intracellular bacteria, such as *Legionella pneumophila*, *Mycobacterium tuberculosis*, *Shigella flexneri*, and *Chlamydia trachomatis*, interfere with mitochondrial dynamics or target other organelles, thus evoking a specific cellular response that could arise as a strategy of the pathogen to ensure its survival, or as a form of defense employed by the host cell to limit infection (Marchi et al., 2022; Tiku, Tan, & Dikic, 2020). In line with this notion, the simultaneous monitoring of both bacteria movement and cellular dynamics in live cells and for extended periods of time (hours) is a crucial aspect for the comprehension of the infection sequence and to detect the alterations that could occur to the entire cellular system. This could be obtained through the HTM-based technology, avoiding phototoxicity and cellular remodeling driven by fluorescence microscopy. For example, HTM has been recently used to detect mitochondrial alterations induced by *Staphylococcus aureus* (Licini et al., 2024), the parasite *Eimeria bovis* (Velasquez, Lopez-Osorio, Mazurek, Hermosilla, & Taubert, 2021), or to understand the subcellular perturbations induced by SARS-CoV-2 infection, revealing drastic generation of syncytia, mitochondrial fragmentation, and accumulation of lipid droplets (Saunders et al., 2024). HTM might be appropriated also to detect changes in cell morphology induced by other perturbators including chemotherapy, RT and metabolic alterations (De Martino, Rathmell, Galluzzi, & Vanpouille-Box, 2024; Galluzzi, Guilbaud, Schmidt, Kroemer, & Marincola, 2024; Yamazaki et al., 2020). It is important to note that not all the intracellular organelles, such as Golgi apparatus and endoplasmic reticulum, can be detected by HTM (Sandoz et al., 2019). Despite these limitations, the HTM approach allows to obtain many information on the morphological rearrangements that occur during pathogen infection, from the bacterial adhesion and entry, its persistence into the host cell, and final exit toward the extracellular milieu, thus revealing potential mechanisms that ensure bacterial survival and proliferation.

Acknowledgments

S.M. is supported by Progetti di Rilevante Interesse Nazionale (PRIN2022ALJN73 and PRINP202243ZAR), and local funds from Marche Polytechnic University (Ancona, Italy). The research leading to these results has received funding from the European Union—NextGenerationEU through the Italian Ministry of University and Research under PNRR—M4C2-I1.3 Project PE_00000019 "HEAL ITALIA" to Saverio Marchi (S.M.), CUP I33C22006900006 (Marche Polytechnic University).

Conflicts of interest

The authors declare that they have no competing interests.

References

Carvalho, F., Spier, A., Chaze, T., Matondo, M., Cossart, P., & Stavru, F. (2020). Listeria monocytogenes exploits mitochondrial contact site and cristae organizing system complex subunit Mic10 to promote mitochondrial fragmentation and cellular infection. *mBio*, *11*, e03171-19.

Charlier, C., Perrodeau, E., Leclercq, A., Cazenave, B., Pilmis, B., Henry, B., et al. (2017). Clinical features and prognostic factors of listeriosis: The MONALISA national prospective cohort study. *The Lancet. Infectious Diseases*, *17*, 510–519.

De Martino, M., Rathmell, J. C., Galluzzi, L., & Vanpouille-Box, C. (2024). Cancer cell metabolism and antitumour immunity. *Nature Reviews. Immunology*, *24*, 654–669.

Galassi, C., Chan, T. A., Vitale, I., & Galluzzi, L. (2024). The hallmarks of cancer immune evasion. *Cancer Cell*, *42*, 1825–1863.

Galluzzi, L., Guilbaud, E., Schmidt, D., Kroemer, G., & Marincola, F. M. (2024). Targeting immunogenic cell stress and death for cancer therapy. *Nature Reviews. Drug Discovery*, *23*, 445–460.

Galluzzi, L., Morselli, E., Vicencio, J. M., Kepp, O., Joza, N., Tajeddine, N., et al. (2008). Life, death and burial: Multifaceted impact of autophagy. *Biochemical Society Transactions*, *36*, 786–790.

Hamon, M. A., Ribet, D., Stavru, F., & Cossart, P. (2012). Listeriolysin O: The Swiss army knife of Listeria. *Trends in Microbiology*, *20*, 360–368.

Kocks, C., Gouin, E., Tabouret, M., Berche, P., Ohayon, H., & Cossart, P. (1992). L. monocytogenes-induced actin assembly requires the actA gene product, a surface protein. *Cell*, *68*, 521–531.

Koopmans, M. M., Brouwer, M. C., Vazquez-Boland, J. A., & van de Beek, D. (2023). Human listeriosis. *Clinical Microbiology Reviews*, *36*, e0006019.

Lebreton, A., Stavru, F., & Cossart, P. (2015). Organelle targeting during bacterial infection: Insights from Listeria. *Trends in Cell Biology*, *25*, 330–338.

Lecuit, M. (2020). Listeria monocytogenes, a model in infection biology. *Cellular Microbiology*, *22*, e13186.

Lecuit, M., Vandormael-Pournin, S., Lefort, J., Huerre, M., Gounon, P., Dupuy, C., et al. (2001). A transgenic model for listeriosis: Role of internalin in crossing the intestinal barrier. *Science*, *292*, 1722–1725.

Li, T., Kong, L., Li, X., Wu, S., Attri, K. S., Li, Y., et al. (2021). Listeria monocytogenes upregulates mitochondrial calcium signalling to inhibit LC3-associated phagocytosis as a survival strategy. *Nature Microbiology*, *6*, 366–379.

Licini, C., Morroni, G., Lucarini, G., Vitto, V. A. M., Orlando, F., Missiroli, S., et al. (2024). ER-mitochondria association negatively affects wound healing by regulating NLRP3 activation. *Cell Death & Disease*, *15*, 407.

Ling, Z., Zhao, D., Xie, X., Yao, H., Wang, Y., Kong, S., et al. (2021). inlF enhances Listeria monocytogenes early-stage infection by inhibiting the inflammatory response. *Frontiers in Cellular and Infection Microbiology*, *11*, 748461.

Marchi, S., Morroni, G., Pinton, P., & Galluzzi, L. (2022). Control of host mitochondria by bacterial pathogens. *Trends in Microbiology*, *30*, 452–465.

Maudet, C., Kheloufi, M., Levallois, S., Gaillard, J., Huang, L., Gaultier, C., et al. (2022). Bacterial inhibition of Fas-mediated killing promotes neuroinvasion and persistence. *Nature*, *603*, 900–906.

Ruan, Y., Rezelj, S., Bedina Zavec, A., Anderluh, G., & Scheuring, S. (2016). Listeriolysin O membrane damaging activity involves arc formation and lineaction—Implication for Listeria monocytogenes escape from phagocytic vacuole. *PLoS Pathogens*, *12*, e1005597.

Sandoz, P. A., Tremblay, C., van der Goot, F. G., & Frechin, M. (2019). Image-based analysis of living mammalian cells using label-free 3D refractive index maps reveals new organelle dynamics and dry mass flux. *PLoS Biology*, *17*, e3000553.

Saunders, N., Monel, B., Cayet, N., Archetti, L., Moreno, H., Jeanne, A., et al. (2024). Dynamic label-free analysis of SARS-CoV-2 infection reveals virus-induced subcellular remodeling. *Nature Communications*, *15*, 4996.

Schindelin, J., Arganda-Carreras, I., Frise, E., Kaynig, V., Longair, M., Pietzsch, T., et al. (2012). Fiji: An open-source platform for biological-image analysis. *Nature Methods*, *9*, 676–682.

Schlech, W. F. (2019). Epidemiology and clinical manifestations of Listeria monocytogenes infection. *Microbiology Spectrum*, 7. https://doi.org/10.1128/microbiolspec.gpp3-0014-2018.

Spier, A., Connor, M. G., Steiner, T., Carvalho, F., Cossart, P., Eisenreich, W., et al. (2021). Mitochondrial respiration restricts Listeria monocytogenes infection by slowing down host cell receptor recycling. *Cell Reports*, *37*, 109989.

Stavru, F., Bouillaud, F., Sartori, A., Ricquier, D., & Cossart, P. (2011). Listeria monocytogenes transiently alters mitochondrial dynamics during infection. *Proceedings of the National Academy of Sciences of the United States of America*, *108*, 3612–3617.

Stavru, F., Palmer, A. E., Wang, C., Youle, R. J., & Cossart, P. (2013). Atypical mitochondrial fission upon bacterial infection. *Proceedings of the National Academy of Sciences of the United States of America*, *110*, 16003–16008.

Tattoli, I., Sorbara, M. T., Yang, C., Tooze, S. A., Philpott, D. J., & Girardin, S. E. (2013). Listeria phospholipases subvert host autophagic defenses by stalling pre-autophagosomal structures. *The EMBO Journal*, *32*, 3066–3078.

Tiku, V., Tan, M. W., & Dikic, I. (2020). Mitochondrial functions in infection and immunity. *Trends in Cell Biology*, *30*, 263–275.

Tilney, L. G., & Portnoy, D. A. (1989). Actin filaments and the growth, movement, and spread of the intracellular bacterial parasite, Listeria monocytogenes. *The Journal of Cell Biology*, *109*, 1597–1608.

Velasquez, Z. D., Lopez-Osorio, S., Mazurek, S., Hermosilla, C., & Taubert, A. (2021). Eimeria bovis macromeront formation induces glycolytic responses and mitochondrial changes in primary host endothelial cells. *Frontiers in Cellular and Infection Microbiology*, *11*, 703413.

Yamazaki, T., Kirchmair, A., Sato, A., Buque, A., Rybstein, M., Petroni, G., et al. (2020). Mitochondrial DNA drives abscopal responses to radiation that are inhibited by autophagy. *Nature Immunology*, *21*, 1160–1171.

Yang, B., Shen, M., Lu, C., Wang, Y., Zhao, X., Zhang, Q., et al. (2024). RNF144A inhibits autophagy by targeting BECN1 for degradation during L. monocytogenes infection. *Autophagy*, 1–18. https://doi.org/10.1080/15548627.2024.2429380.

Zhang, Y., Yao, Y., Qiu, X., Wang, G., Hu, Z., Chen, S., et al. (2019). Listeria hijacks host mitophagy through a novel mitophagy receptor to evade killing. *Nature Immunology*, *20*, 433–446.

CHAPTER EIGHT

A screening system to determine the effect of bacterial metabolites on MAdCAM-1 expression by transformed endothelial sinusoidal cells

Ai-Ling Tian[a,b], Marion Leduc[a,b], Marine Fidelle[c,d], Laurence Zitvogel[c,d,e,f], Guido Kroemer[a,b,g,*], and Oliver Kepp[a,b,*]

[a]Metabolomics and Cell Biology Platforms, Gustave Roussy Cancer Center, Université Paris Saclay, Villejuif, France
[b]Centre de Recherche des Cordeliers, Equipe labellisée par la Ligue contre le cancer, Université Paris Cité, Sorbonne Université, INSERM U1138, Institut Universitaire de France, Paris, France
[c]Gustave Roussy, Villejuif, France
[d]INSERM U1015, Equipe Labellisée—Ligue Nationale contre le Cancer, Villejuif, France
[e]Faculté de Médecine, Université Paris-Saclay, Kremlin-Bicêtre, France
[f]Center of Clinical Investigations for In Situ Biotherapies of Cancer (BIOTHERIS) INSERM, CIC1428, Villejuif, France
[g]Institut du Cancer Paris CARPEM, Department of Biology, APHP, Hôpital Européen Georges Pompidou, Paris, France
*Corresponding authors: e-mail address: kroemer@orange.fr; captain.olsen@gmail.com

Contents

Methods in Cell Biology, Volume 194
ISSN 0091-679X
https://doi.org/10.1016/bs.mcb.2024.01.007

Abstract

Mucosal addressin cell adhesion molecule 1 (MAdCAM-1) expression in high endothelial venules is regulated by bacterial metabolites emanating from the gut and the interaction of MAdCAM-1 with α4β7 integrin mediates lymphocyte diapedesis into gut-associated secondary lymphoid tissues. MAdCAM-1 thus controls the abundance of circulating immunosuppressive T cells that can reach malignant tissue and compromise the therapeutic efficacy of anticancer immunotherapy. Here we describe a biosensor-based phenotypic assessment that facilitates the high throughput screening (HTS)-compatible assessment of MAdCAM-1 regulation in response to exposure to bacterial metabolites. This screening routine encompasses high endothelial venule cells expressing green fluorescent protein (GFP) under the control of the MAdCAM-1 promoter combined with robot-assisted bioimaging and a multistep image analysis pipeline. Altogether this system facilitates the discovery of bacterial composites that control anticancer immunity via the sequestration of Th17-specific regulatory T cells (Treg17) in the gut.

1. Introduction

The composition of the gut microbiome plays a crucial role in innate and adaptive immunity as it maintains the homeostatic function of gut regulatory T (Treg) cells, thus impacting the immune response to pathological insults and malignant transformation (Finlay et al., 2020). In consequence, intestinal dysbiosis has been recognized as general feature of declining health and is often associated with advanced aging (Lopez-Otin & Kroemer, 2021; Lopez-Otin, Pietrocola, Roiz-Valle, Galluzzi, & Kroemer, 2023). Gut microbial dysbiosis, which can be caused by the use of broad-spectrum antibiotics, is also associated with the lack of efficacy of anticancer immunotherapy across various types of malignancies not restricted gastrointestinal cancers (Fidelle, Tian, Zitvogel, & Kroemer, 2023). The concomitant use of broad-spectrum antibiotics with immune checkpoint inhibitors (ICIs) targeting the PD-1/PD-L1 axis has been shown to reduce the therapeutic efficacy of the cancer therapy and is associated with disease progression and poor outcome (Derosa et al., 2021). Mechanistically, both in mice and in cancer patients, gut dysbiosis resulting from the treatment with broad-spectrum antibiotics arises in part due to the recolonization of the gut by *Enterocloster* spp. (i.e., *E. clostridioformis*, *E. bolteae*), which negatively affects the outcome of immunotherapy (Derosa et al., 2021; Fidelle, Tian, et al., 2023).

In this setting, bacterial composites trigger reprogramming of the tumor microenvironment (TME) via the regulation of mucosal addressin cell adhesion molecule 1 (MAdCAM-1) expression on high endothelial venules

(HEV) in the gut. The interaction of MAdCAM-1 with α4β7 integrins expressed on T cells increases the adhesion of lymphocytes to HEV and their consequent diapedesis across the vascular endothelial barrier to gut-associated secondary lymphoid tissue (GALT). MAdCAM-1 is constitutively expressed in high HEV. However, the level of expression of MAdCAM-1 can be regulated by inflammatory cytokines and bacterial metabolites (Ogawa et al., 2005). The recolonization following antibiotic treatment of the gut by *E. clostridioformis* impacts the metabolism of bile acids, which in turn leads to a downregulation of MAdCAM-1 expression (Fidelle, Rauber, et al., 2023). The resulting exodus of intestinal immunosuppressive Th17-specific regulatory T cells (Treg17) cells toward the tumor microenvironment then interferes with T cell-targeted immunotherapy. Altogether, the MAdCAM-1-α4β7 interaction keeps Treg17 cells from migrating from the gut to tumors. For this reason, gut dysbiosis, which decreases MAdCAM-1 expression, compromises cancer immunosurveillance, as well as the anticancer effects of ICIs.

Here we generated a biological platform for the high throughput screening (HTS)-compatible assessment of MAdCAM-1 expression that allows to screen bacterial metabolites as well as pharmacological agents for their effects on the MAdCAM-1-α4β7 axis. This tool can be used for proof-of-concept studies identifying potentially harmful bacterial composites. In addition, this method can be upscaled to high-throughput screening campaigns that may lead to the identification of drugs that counteract the adverse effects of gut dysbiosis.

2. Materials and methods

2.1 Cell lines and culture reagents

Culture media and supplements for cell culture were obtained from Life Technologies (Carlsbad, CA, USA) and plastic material came from Greiner Bio-One (Kremsmünster, Austria) and Corning (Corning, NY, USA). Mouse transformed endothelial sinusoidal cells (TSECs) were a generous gift from Professor B.P. Fennimore (University of Colorado, Aurora, CO, USA). The medium for this cell line is Dulbecco's modified Eagle medium/nutrient mixture F-12 (DMEM/F-12, Catalog No. 11320074, Gibco).

To complement growth medium, following components are added to the base medium:

- fetal bovine serum (FBS) to a final concentration of 10%,
- endothelial cell growth supplement (ECGS, Catalog No. 1052, ScienCell) diluted to 1×,
- 10 U/mL of penicillin sodium and 10 μg/mL of streptomycin sulfate.

TSECs were cultured in an air atmosphere enriched with 5% CO_2 at 37 °C.

2.2 Cell culture procedures

1. Thaw the cells by gentle agitation of the cryovial in a 37 °C water bath. To reduce the possibility of contamination, keep the O-ring and cap from contacting the surface of the water. Thawing should be rapid (approximately 2 min).
2. Remove the vial from the water bath as soon as the contents are thawed and decontaminate by dipping in or spraying with 70% ethanol. All steps from this point on should be carried out under strict aseptic conditions.
3. Transfer the vial's content to a centrifuge tube containing 9.0 mL complete growth medium, and spin at approximately 125 × *g* for 5–7 min.
4. Resuspend the cell pellet with complete growth medium. It is important to avoid excessive alkalinity of the medium during the recovery of cells. It is suggested that, prior to the addition of the cells, the culture vessel containing the complete growth medium is placed in a cell culture incubator for at least 15 min to allow the medium to stabilize pH (7.0–7.6).
5. Incubate the culture in an air atmosphere enriched with 5% CO_2 at 37 °C.

2.3 Sub-culture procedures

1. Remove and discard the culture medium.
2. Briefly rinse the cell layer with 0.25% (w/v) Trypsin-0.03% (w/v) EDTA solution to remove all traces of serum which contain trypsin inhibitor.
3. Add 1–2 mL of Trypsin-EDTA solution to the flask and observe cells under an inverted microscope until the cell layer begins to detach.
4. Add 6–8 mL of complete growth medium and aspirate cells by gently pipetting.
5. Add appropriate aliquots of cell suspension to new culture vessels.
6. Every 2–3 days renew the medium.

2.4 Establishment of Geneticin™ selection condition

1. Plate TSECs at a density of 2×10^4 cells/mL in 0.5 mL complete growth medium per well in a 24-well tissue culture plate.
2. Culture overnight. Most cell types should be ≥80% confluent prior to adding the selection antibiotic.

3. Add increasing amounts of Geneticin™ selection antibiotic (G418 sulfate) (Catalog No. 10131027, Gibco) to duplicate wells of cells plated in complete growth medium. We suggest to use 0.1, 0.2, 0.3, 0.4, 0.5, 0.6, 0.8, 1.0, 1.2, 1.5, and 2.0 mg/mL for establishing a killing curve and for identifying the optimal concentration of G418 sulfate for a given cell type. Include a control not containing selection antibiotic.
4. Replace the medium containing selection antibiotic every 2–3 days for up to a week. Examine the culture every day for signs of visual toxicity. Determine the optimal antibiotic dose which corresponds to the lowest antibiotic concentration at which all wild type cells die within 7 days of selection.
5. Proceed with stable cell line generation using the concentration determined in step 4 (in our case 0.5 mg/mL).

2.5 MAdCAM-1 promoter plasmid

2.5.1 Preparation of linearized vectors

1. Select appropriate cloning site to linearize pAcGFP1-1 vector (Catalog No. PT3846-5, Takara).
2. The linearized vector can be obtained by digesting the circular vector with *Eco*RI (Catalog No. R3101, NEB) and *Sac*II (Catalog No. R0157, NEB). Set up reaction as follows:

Component	50 μL reaction
DNA	1 μg
10× rCutSmart buffer	5 μL (1×)
*Eco*RI-HF	1.0 μL (20 units)
*Sac*II	1.0 μL (20 units)
Nuclease-free water	to 50 μL

3. Incubate at 37 °C overnight.

2.5.2 Preparation of inserts

1. The general principle for the primer design: insertion of a homologous sequence (15–20 bp, excluding restriction sites) into the 5′ ends of both

forward and reverse primers. Thereby, the ends of amplified insert and linearized vectors are complementary.

- Forward primer: 5′tcgagctcaagcttcgaattcTACCCCCACAGGCC TGCC-3′
- Reverse primer: 5′cggtggatcccgggcccgcggGGGCCGGCAGCTTC CTAC-3′

2. PCR amplification of MAdCAM-1 promoter sequence. Reaction setup:

Component	50 μL reaction	Final concentration
10 μM forward primer	1 μL	0.2 μM
10 μM reverse primer	1 μL	0.2 μM
Template DNA[a]	1 μg	1 μg
OneTaq® Hot Start Quick-Load® 2× Master Mix with Standard Buffer (Catalog No. M0488, NEB)	25 μL	1×
Nuclease-free water	to 50 μL	

[a]Template DNA: mMAdCAM-1 Promotor_pUC57-Kan (Catalog No. U135WGI270, synthesized by GenScript).

3. Transfer PCR tubes to a PCR machine and begin thermocycling. Thermocycling conditions for a routine PCR:

Step	Temperature	Time
Initial denaturation	94 °C	30 s
30 cycles of denaturation, annealing, extension	94 °C	30 s
	68 °C	60 s
	68 °C	2 min
Final extension	68 °C	5 min
Hold	4 °C	

2.5.3 Ratio of vector to insert

For ClonExpress MultiS recombination reaction system, the optimal amount is 0.03 pmol per fragments (including linearized vector). This mass can be roughly calculated according to the following formula:

- The optimal mass of vector required = [0.02 × number of base pairs] ng (0.03 pmol)

For example, our cloning inserts of 2 kb into the pAcGFP1-1 vector of 4.1 kb, the optimal mass of the vector and insert are as follows:

- The optimal mass of linearized vector required: 0.02 × 4100 = 82 ng
- The optimal mass of insert of 2 kb required: 0.02 × 2000 = 40 ng

2.5.4 Recombination

1. The amount of DNA can be roughly calculated according to the above formula.
2. Dilute the pAcGFP1-1 vector and insert at an appropriate ratio to ensure the accuracy of pipetting before preparing the recombination reaction system, and the amount of each component is not less than 1 μL.
3. Prepare the following reaction on ice by using ClonExpress® MultiS One Step Cloning Kit (Catalog No. C113, Vazyme):

Components	Recombination	Negative control-1[a]	Negative control-2[b]	Positive control[c]
Linearized vector[d]	X μL	X μL	0 μL	1 μL
Insert[d]	Y_1 Y_n μL	0 μL	Y_1 Y_n μL	1 μL
5× CE MultiS buffer	4 μL	0 μL	0 μL	4 μL
Exnase MultiS	2 μL	0 μL	0 μL	2 μL
ddH_2O	to 20 μL	to 20 μL	to 20 μL	to 20 μL

[a]It is recommended to use negative control-1, which can confirm the residue of the cyclic plasmid template.
[b]It is recommended to use negative control-2, when the amplification template of the insert is a circular plasmid with the same antibiotic resistance as the cloning vector. It is recommended to perform the circular plasmid residue detection of the linearized vector and inserts independently.
[c]It is recommended to include positive controls to exclude technical problems.
[d]X/Y is the amount of calculated vector/insert ratio.

4. Gently pipette up and down several times to mix thoroughly (DO NOT VORTEX!). Briefly centrifuge to collect the reaction mix to the bottom of the tube.
5. Incubate at 37 °C for 30 min and immediately chill the tube at 4 °C or on ice.

2.5.5 High efficiency transformation

1. Thaw a tube of NEB 5-alpha competent *E. coli* cells (Catalog No. C2987H, NEB) on ice for 10 min.
2. Pipette 5 μL of the recombination products to the cell mixture.
3. Carefully flick the tube wall 4–5 times to mix thoroughly (DO NOT VORTEX!).
4. Place the mixture on ice for 30 min.
5. Heat shock at exactly 42 °C for exactly 45 s and then immediately place on ice for 5 min.
6. Pipette 950 μL of room temperature SOC liquid medium (without antibiotics) into the mixture.
7. Shake vigorously at 37 °C for 60 min at 250 rpm.
8. Preheat the kanamycin (50 μg/mL) resistant LB solid medium plates in a 37 incubator.
9. Centrifuge the culture at 2400 × *g* for 5 min and discard 900 μL of supernatant.
10. Use the remaining medium to suspend the bacteria.
11. Use a sterile bent glass rod to gently spread onto the plate containing kanamycin (50 μg/mL) resistant.
12. Incubate at 37 °C for 12–16 h.

2.5.6 Clone picking and DNA purification

1. After overnight incubation, hundreds of single clones will form on the plate of recombination reaction, whereas none of those will grow on the plate of the negative control(s).
2. Pick several clones from the plate of recombination reaction for colony PCR with at least one universal sequencing primer of the vector. If the colony is positive, a band slightly larger than the size of the inserts should appear.
3. Inoculate the remaining bacterial solution of positive clones into liquid LB medium containing kanamycin (50 μg/mL) for overnight incubation.

4. Purify the plasmid using the Monarch® plasmid miniprep kit (Catalog No. T1010, NEB) for restriction endonuclease digestion identification, and perform DNA sequencing to check for the correct insertion of the MAdCAM-1 promotor at the expected *Eco*RI and *Sac*II restriction site in the pAcGFP1-1 plasmid using the following primers:

Name	Sequence 5′ to 3′
mSeqPrimer1	GATTCTGTGGATAACCGTATTACC
mSeqPrimer2	CCTGTACATCCAGTCAGCAC
mSeqPrimer3	AGATTTCCATGCACTTGACC

5. Identify the correct pAcGFP1-1 construct and grow in ~200 mL of liquid LB medium containing kanamycin (50 μg/mL) resistant in order to prepare a larger amount of DNA.
6. Purify the plasmid DNA using the PureLink™ expi endotoxin-free maxi plasmid purification kit (Catalog No. A31231, Invitrogen) and determine the concentration using a Nanodrop device (Thermo Fisher Scientific).

2.6 MAdCAM-1 promoter reporter cell line

2.6.1 Cell seeding

1. To prepare cells, collect enough cells to complete the transfection experiment, and centrifuge for 5 min at $300 \times g$ in a swing-bucket rotor. Suspend the cell pellet to an appropriate concentration in medium.
2. Plate 0.5×10^5 TSECs in 1 mL of complete growth medium per well of a 6-well plate.

2.6.2 Preparing the FuGENE® HD transfection reagent

1. Before use, allow the vial of FuGENE® HD transfection reagent (Catalog No. E2311, Promega) to reach room temperature.
2. Mix by inverting or vortexing briefly. No precipitate should be visible.

2.6.3 Transfection

1. The total volume of medium, DNA, and FuGENE® HD transfection reagent to add per well of a 6-well plate is 80 μL. To a sterile tube,

add 920 μL of complete growth medium prewarmed to room temperature so that the final volume after addition of DNA is 1 mL. Add 20 μg of plasmid DNA (1 μg/μL) to the medium, and vortex. For a 3:1 FuGENE® HD transfection reagent:DNA ratio, add 60 μL of FuGENE® HD transfection reagent, and mix immediately.

2. Incubate the FuGENE® HD transfection reagent/DNA mixture for 15 min at room temperature.
3. Add 80 μL of the FuGENE® HD transfection reagent/DNA mixture per well to the 6-well plate containing 1 mL of cells in the complete growth medium. Mix well by shaking.
4. Return cells to an air atmosphere enriched with 5% CO_2 at 37 °C for 48 h.
5. Forty-eight hours post-transfection, harvest adherent cells and plate at several different dilutions (e.g., 1:2, 1:5, 1:10) in selection medium (complete growth medium containing 0.5 mg/mL of G418).
6. For the next 14 days, replace the selection medium every 3–4 days.
7. During the second week, monitor cells for distinct the formation of colonies. Complete cell death should occur in cultures transfected with the negative control plasmid.
8. The G418-resistant cells are sorted into single-cell clones in 96-well plates (one cell per well) to establish a TSEC::MAdCAM-1-GFP stable cell line expressing GFP under the control of the MAdCAM-1 promoter.

2.6.4 Identification of positive clones

1. Once the clones start forming colonies visible by light microscopy, the exhausted medium is removed and 50 μL of Trypsin-EDTA solution is dropped into the well to detach the cells. Trypsin-mediated digestion is stopped by the addition of 150 μL fresh selective medium, and the cellular suspension is transferred to a new 96-well plate.
2. The plate that holds the selected clones is duplicated and one plate is kept as a maintenance culture whereas the other is cultured under the stimulation of recombinant murine TNF-α (20 ng/mL, Catalog No. 315-01A, PeproTech) and IL-1β (20 ng/mL, Catalog No. 211-11B, PeproTech) in the presence of 5% CO_2 at 37 °C for 48 h.
3. Clones that homogeneously express GFP are select based on fluorescence microscopic assessment compared with the same but non-treated control clones. Selected clones are named TSEC::MAdCAM-1-GFP cells (Fig. 2A).

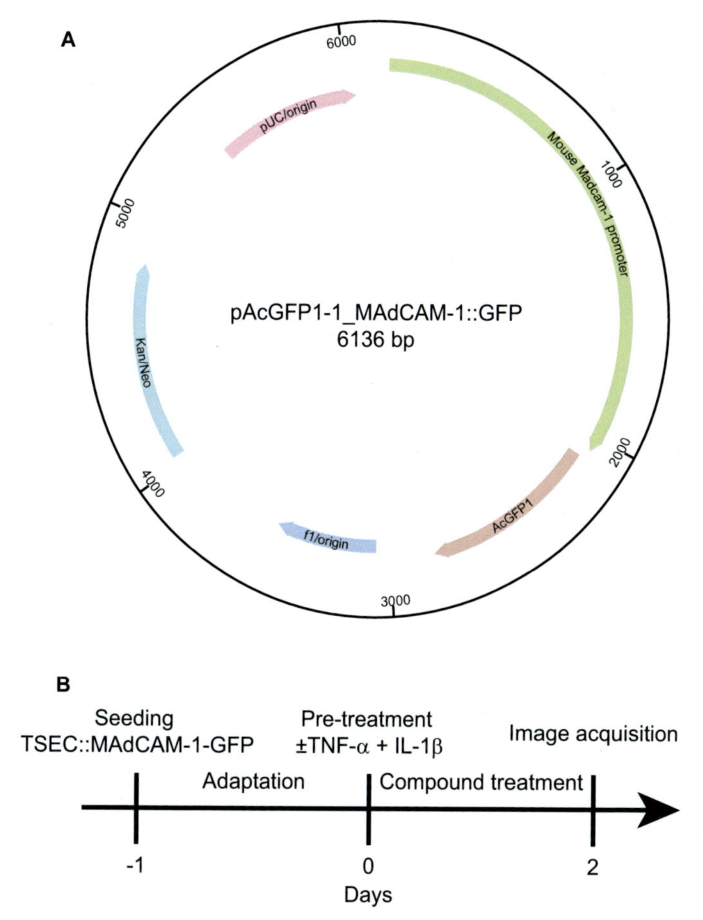

Fig. 1 Plasmid map and treatment scheme. The MAdCAM-1-GFP promoter reporter plasmid (A) was used to generate TSEC::MAdCAM-1-GFP cells, which then were employed for high throughput screening campaigns according to the depicted schedule (B).

2.6.5 Treatment with bacterial metabolites *(Fig. 1B)*

1. Recombinant murine TNF-α and IL-1β were diluted in culture media from a stock of 40 μg/mL at the time of treatment. Bile acids (lithocholic acid (LCA) (Catalog No. 700218P, Sigma-Aldrich), sodium glycodeoxycholate (GDCA) (Catalog No. G9910, Sigma-Aldrich), sodium chenodeoxycholate (CDCA) (Catalog No. C8261, Sigma-Aldrich), deoxycholic acid (DCA) (Catalog No. 30960, Sigma-Aldrich), tauroursodeoxycholic acid (TUDCA) (Catalog No. 580549, Sigma-Aldrich), glycoursodeoxycholic acid (GUDCA) (Catalog No. 06863,

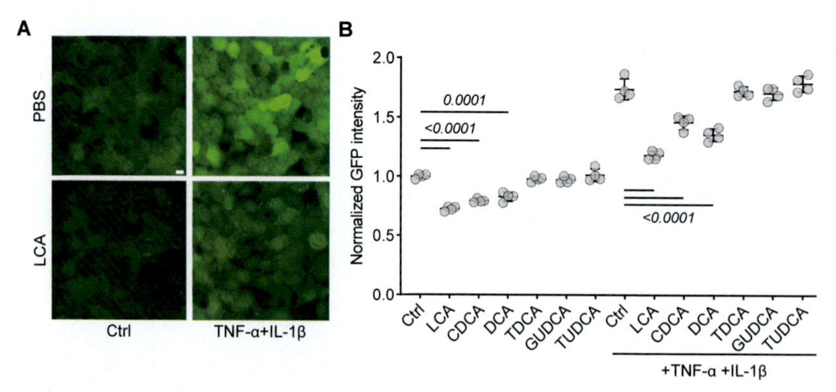

Fig. 2 Effect of *Enterocloster clostridioformis*-induced perturbation of ileal biliary acids on MAdCAM-1 expression. TSEC::MAdCAM-1-GFP cells exposed to multiple bile acids at a final concentration of 100 μM were used in high content confocal microscopy. Representative images are shown in (A). Scale bar equals 10 μm. Results are depicted as normalized cell fluorescent intensity of GFP (B). Mean ± SD of triplicates from four independent experiments. Comparisons between groups were analyzed using one-way ANOVA.

Sigma-Aldrich), taurodeoxycholic acid (TDCA) (Catalog No. 580221, Sigma-Aldrich)) dissolved in DMSO were diluted 1:1000 in culture media to a final concentration of 100 μM.

2. TSEC::MAdCAM-1-GFP cells were seeded in 96-well μclear imaging plates (Greiner Bio-One) and allowed to adapt overnight.
3. Cells were treated with the indicated agents for 48 h at 37 °C with 5% CO_2, then fixed with 3.7% paraformaldehyde (PFA, w/v in PBS) (Catalog No. F8775, Sigma-Aldrich) containing 2 μg/mL of Hoechst 33342 (Catalog No. H3570, Invitrogen) at 4 °C overnight.
4. Subsequently, the fixative was exchanged with PBS for the further microscope scanning (Fig. 2).

2.7 Image acquisition and processing

Fixed plates underwent comprehensive analysis through automated microscopy. Image acquisition was conducted employing the ImageXpress Micro C automated confocal microscope (Molecular Devices, Sunnyvale, CA, USA), equipped with a 20× PlanApo objective lens (Nikon, Tokyo, Japan). Subsequently, automated image processing was executed utilizing the R software, with both the EBImage package (accessible from the Bioconductor repository at https://www.bioconductor.org) and the MorphR package (available from the GitHub repository at https://github.com/kroemerlab/MorphR). Normalization of images was carried out on a plate-wise basis, employing the pixel intensity distribution of control samples. This normalization procedure aimed to eliminate extreme intensities and enhance image contrast. Nuclei were detected utilizing a combination of adaptive thresholding, followed by dimension reduction through maximum projection, succeeded by three morphological operations to refine the mask, and finally, a watershed transformation to segment individual objects. For FITC images, a normalization process was applied within an automatically computed range of intensities to enhance cytoplasmic signal. Cytoplasmic masks were generated through a sigmoid transformation in conjunction with Otsu thresholding. Subsequent morphological operations were applied to further refine the cytoplasmic mask, resulting in object labeling. The extraction of cellular parameters was performed utilizing distinct masks.

2.8 Data analysis

After exclusion of cellular debris and non-viable cells, the data underwent normalization, followed by rigorous statistical evaluation and graphical representation. Also using R software, representative images were extracted and pixel intensities scaled (to the same extent for all images of a given experiment). A minimum of four distinct view fields were systematically analyzed per well, and each experimental assessment was performed in quadruplicate. The results obtained from the wells were subjected to comprehensive statistical analysis employing one-way Mann-Whitney *U* test and were then graphically displayed using GraphPad Prism (Marchi, Morroni, Pinton, & Galluzzi, 2022).

3. Concluding remarks

The interplay between bacteria and the host organism has been largely explored (Marchi et al., 2022; Rudel, Kepp, & Kozjak-Pavlovic, 2010; Tiku, Tan, & Dikic, 2020). Nevertheless, the mechanistic exploration of gut microbial dysbiosis is limited by the lack of high-throughput assays compatible with large-scale screening efforts in appropriate cells types. Here we developed a promoter reporter system in high endothelial venule cells that allows for an automation friendly assessment of MAdCAM-1 expression. Combined with robotized workflows, image-based analysis routines and standardized data processing the described platform allows for the identification of bacterial metabolites and pharmacological agents that reduce or enhance MAdCAM-1 expression. Metabolites and agents identified with this routine can impact innate and adaptive immunity and might thus be of translational value in the area of immuno-oncology.

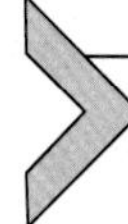

4. Notes

1. Replace cell culture medium every 2–3 days. Cells should be passaged after reaching 70–80% confluency.
2. Corning® T-75 flasks (Catalog No. 430641) are recommended for the culture of TSECs. A subcultivation ratio of 1:50 to 1:100 is recommended.
3. Gently mix the PCR amplification reaction. Collect all liquid to the bottom of the tube by a quick spin if necessary. Overlay the sample with mineral oil if using a PCR machine without a heated lid.
4. NEB 5-alpha Competent *E. coli* cells are best thawed on ice and DNA is added as soon as the last ice in the tube disappears. Cells can also be thawed by hand, but warming above 0 °C will decrease the transformation efficiency.
5. The volume of recombination products when needs high efficiency transformation should be $\leq 1/10$ of the volume of competent cells.
6. Do not use cells after passage 9–10 as this may lead to reduced viabilities and transfection efficiencies.
7. Cells should be passaged 2–5 days before transfection depending on growth rate of cells.
8. Plate adherent cells 1 day before transfection so that cells are approximately 80% confluent on the day of transfection.
9. Cells transfected with a plasmid harboring the neomycin resistance (*neo*) gene should be grown in the complete growth medium for 48–72 h post-transfection before the selection antibiotic is applied.
10. Image quality of fluorescent intensity in the GFP channel can be ameliorated by using a maximum projection of a Z-stack.
11. The exposure time is adjusted according to the signal intensity of each channel.

Disclosures

O.K. and G.K. have been holding research contracts with Daiichi Sankyo, Eleor, Kaleido, Lytix Pharma, PharmaMar, Osasuna Therapeutics, Samsara Therapeutics, Sanofi, Tollys, and Vascage. G.K. is on the Board of Directors of the Bristol Myers Squibb Foundation France. G.K. is a scientific co-founder of everImmune, Osasuna Therapeutics, Samsara Therapeutics and Therafast Bio. O.K. is a scientific co-founder of Samsara Therapeutics. G.K. is in the scientific advisory boards of Hevolution, Institut Servier and Longevity Vision Funds. G.K. is the inventor of patents covering therapeutic targeting of aging, cancer, cystic fibrosis and metabolic disorders. G.K.'s brother, Romano Kroemer, was an employee of Sanofi and now consults for Boehringer-Ingelheim.

Acknowledgments

M.F. is supported by the SEERAVE Foundation, O.K. receives funding from Institut National du Cancer (INCa); G.K. is supported by the Ligue contre le Cancer (équipe labellisée); Agence National de la Recherche (ANR)—Projets blancs; AMMICa US23/CNRS UMS3655; Association pour la recherche sur le cancer (ARC); Cancéropôle Ile-de-France; Fondation pour la Recherche Médicale (FRM); a donation by Elior; Equipex Onco-Pheno-Screen; European Joint Programme on Rare Diseases (EJPRD); European Research Council (ICD-Cancer), European Union Horizon 2020 Projects Oncobiome and Crimson; Fondation Carrefour; INCa; Institut Universitaire de France; LabEx Immuno-Oncology (ANR-18-IDEX-0001); a Cancer Research ASPIRE Award from the Mark Foundation; the RHU Immunolife; Seerave Foundation; SIRIC Stratified Oncology Cell DNA Repair and Tumor Immune Elimination (SOCRATE); and SIRIC Cancer Research and Personalized Medicine (CARPEM). This study contributes to the IdEx Université de Paris ANR-18-IDEX-0001.

References

Derosa, L., Routy, B., Desilets, A., Daillere, R., Terrisse, S., Kroemer, G., et al. (2021). Microbiota-centered interventions: The next breakthrough in immuno-oncology? *Cancer Discovery*, *11*(10), 2396–2412.

Fidelle, M., Rauber, C., Alves Costa Silva, C., Tian, A. L., Lahmar, I., de La Varende, A. M., et al. (2023). A microbiota-modulated checkpoint directs immunosuppressive intestinal T cells into cancers. *Science*, *380*(6649), eabo2296.

Fidelle, M., Tian, A. L., Zitvogel, L., & Kroemer, G. (2023). Bile acids regulate MAdCAM-1 expression to link the gut microbiota to cancer immunosurveillance. *Oncoimmunology*, *12*(1), 2224672.

Finlay, B. B., Goldszmid, R., Honda, K., Trinchieri, G., Wargo, J., & Zitvogel, L. (2020). Can we harness the microbiota to enhance the efficacy of cancer immunotherapy? *Nature Reviews. Immunology*, *20*(9), 522–528.

Lopez-Otin, C., & Kroemer, G. (2021). Hallmarks of health. *Cell*, *184*(7), 1929–1939.

Lopez-Otin, C., Pietrocola, F., Roiz-Valle, D., Galluzzi, L., & Kroemer, G. (2023). Meta-hallmarks of aging and cancer. *Cell Metabolism*, *35*(1), 12–35.

Marchi, S., Morroni, G., Pinton, P., & Galluzzi, L. (2022). Control of host mitochondria by bacterial pathogens. *Trends in Microbiology*, *30*(5), 452–465.

Ogawa, H., Binion, D. G., Heidemann, J., Theriot, M., Fisher, P. J., Johnson, N. A., et al. (2005). Mechanisms of MAdCAM-1 gene expression in human intestinal microvascular endothelial cells. *American Journal of Physiology. Cell Physiology*, *288*(2), C272–C281.

Rudel, T., Kepp, O., & Kozjak-Pavlovic, V. (2010). Interactions between bacterial pathogens and mitochondrial cell death pathways. *Nature Reviews. Microbiology*, *8*(10), 693–705.

Tiku, V., Tan, M. W., & Dikic, I. (2020). Mitochondrial functions in infection and immunity. *Trends in Cell Biology*, *30*(4), 263–275.

CHAPTER NINE

Monitoring the mitochondrial localization of mycobacterial proteins

Krishnaveni Mohareer*, Jayashankar Medikonda†, Sriram Yandrapally, Anushka Agarwal, and Sharmistha Banerjee*

Department of Biochemistry, School of Life Sciences, University of Hyderabad, India

*Corresponding authors: e-mail address: krishnaveni.mohareer@gmail.com; sbsl@uohyd.ac.in

Contents

† Present address: Research and Development, Aurovaccine pvt limited, Indrakaran, Sangareddy, Telangana, India 502329.

Methods in Cell Biology, Volume 194

ISSN 0091-679X

https://doi.org/10.1016/bs.mcb.2024.10.018

Abstract

Mitochondrion apart from being the energy hub of the cell, is also the center of various signaling pathways. During intracellular infection, either bacterial or viral, several pathogen proteins, metabolites, and possibly, lipids interact with the host mitochondria. These interactions allow the pathogens, such as *Mtb*, to reprogram the host mitochondrial functions, screwing the host immune responses and resulting in the persistence of the bacteria. Therefore, mitochondria offer a critical target organelle for various therapeutic interventions. This chapter deals with methods to demonstrate and establish the mitochondrial localization of *Mtb* proteins by confocal microscopy and mitochondrial enrichment. Transient transfection of mammalian constructs or infection *Mycolicibacterium smegmatis* infection could be used to overexpress the candidate *Mtb* protein in host cells, allowing the study of changes in the mitochondria's composition and function with regard to localization studies, mitochondrial DNA and RNA, proteomics, metabolomics, lipidomics, and bioenergetics.

1. Introduction

Several host-pathogen interactions, such as bacterial or viral infections, have been demonstrated to be successful owing to the hijack of the host mitochondria, which are the hub of host cell signaling and innate immune responses (Elesela & Lukacs, 2021; Marchi, Morroni, Pinton, & Galluzzi, 2022; Tiku, Tan, & Dikic, 2020). The pathogens affect the host mitochondrial metabolism leading to altered innate signaling and functions. These alterations cause mitochondrial dysfunction leading to accumulation of toxic products affecting cellular homeostasis such as Ca^{2+} fluxes, bioenergetics, mitochondrial dynamics, and cell fate that underline numerous pathological processes (Behl et al., 2023). Reversal of mitochondrial dysfunction by small molecules (Asalla, Mohareer, & Banerjee, 2017; Cartes-Velásquez et al., 2024) or siRNA (Gao et al., 2021; Lum & Morona, 2014) leads to pathogen clearance. Therefore, understanding of the mitochondrial biology during host-pathogen interaction is crucial for the development of intervention strategies against both infectious diseases and other pathologies.

For the present chapter, we focus on by *Mycobacterium tuberculosis* (*Mtb*) infection of human host cells, wherein, mitochondria are often subjected to manipulation in terms of structure and function leading to changes in the host environment that favor the survival of the pathogen (Asalla et al., 2017; Brokatzky & Häcker, 2022; Jamwal et al., 2013). Mycobacterial infection of macrophages causes remarkable changes in mitochondrial dynamics and energy metabolism that ultimately change the immunometabolic state and bacterial persistence (Kumar et al., 2019; Mohareer & Banerjee, 2023; Mohareer, Medikonda, Vadankula, & Banerjee, 2020). The rounded or punctuate mitochondria found in *Mtb*-infected macrophages are caused by altered mitochondrial dynamics with increased fragmentation (Asalla et al., 2017). The altered mitochondrial morphology further impacts its bioenergetics and metabolism that spirals into the regulation of innate immune signaling and thus cell fate (Mohareer et al., 2020).

These alterations in mitochondria are driven by proteins and metabolites secreted by *Mtb*. Systematic bioinformatics analyses have shown that around 1% of the genome of MAP (*Mycobacterium avium* subspecies *paratuberculosis*) contributes to proteins that potentially target host mitochondria (Rana, Rub, & Akhter, 2014). The mitochondrial targeting proteins either carry the conventional MTS (mitochondrial targeting sequence) or satisfy positive charge requirements for mitochondrial localization such as Rv1818c/PE_PGRS33 (Cadieux et al., 2011), Rv0475 (Sohn et al., 2011) and Rv0547c ΔMTS (Medikonda et al., 2024). The targeting protein reaches one of the different mitochondrial sub-compartments, such as the outer mitochondrial membrane, the inner mitochondrial membrane, or the intermembrane space. Based on the sub-compartment, the targeting protein may influence diverse signaling or host mitochondrial physiology. Some of the altered host functions include oxidative phosphorylation (Dubey et al., 2021), induce cell death modalities such as apoptosis (Cadieux et al., 2011) or necrosis (Tundup, Mohareer, & Hasnain, 2014), or potentially inhibit cell death (Joseph, Yuen, Singh, & Hmama, 2017). In the case of *Mtb*, several of its secreted proteins are known to enter mitochondria and alter its bioenergetics or signaling and also bind to mitochondrial DNA, affecting the transcription of oxidative phosphorylation subunits. However, a detailed investigation of sub-mitochondrial localization of *Mtb* proteins is lacking. In most cases, the targeting proteins disrupt mitochondrial membrane potential and induce cell death (Mohareer, Asalla, & Banerjee, 2018), while in some cases, cause inhibition of cell death and sometimes alter membrane fluidity (Medikonda et al., 2024). There occurs a tight regulation of the different host cell death modalities (Mohareer et al., 2018) and therefore the factors that are involved in the execution of the same are also tightly regulated to maintain cellular homeostasis during the host-pathogen interaction.

This chapter helps the reader to gain insights into experimental techniques that help us understand if a secreted protein of a pathogen (or others) is indeed targeted to mitochondria. A detailed investigation of sub-mitochondrial localization is advised through super resolution and transmission electron microscopy (not covered in the chapter) to gain functional insights into the role and level of pathogen interference. Mitochondrial targeting of a given candidate protein can be evaluated for different regions, residues, alternate MTS tags, and so on to evaluate the physiological significance of mitochondrial targeting using the methods outlined in this chapter (also others) and establish the molecular basis of mitochondrial targeting. Subsequently, the researcher could generate mutants by sequence analysis and further mutating the predicted mitochondrial targeting residues. These mutants will further help in understanding the implications or consequences of mitochondrial targeting and alterations in the host functioning or cell fate. Besides, the mitochondria-enriched fractions can be used for biochemical validation and various other downstream applications such as proteomics by MS-MS, lipidomics by LC-MS, metabolomics by GC–MS, and mitochondrial complex assays that would add more information on the mechanistic details of their functional relevance in host-pathogen interaction.

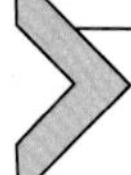

2. Materials

2.1 Disposables

- ice
- 10, 200 and 1000 μL Universal Pipet Tips (#521000, #521010, #521020, Tarsons)[a]
- 1.5 mL Micro centrifuge tubes, (#500010, Tarsons)
- Eppendorf Serological pipet, 10 mL, 5 mL, individually wrapped, orange, (# 0030127.722, # 0030127714)
- Eppendorf Cell Culture Plate, 12-well, TC treated, individually wrapped, (#0030721.110)
- Eppendorf Cell Culture dish, 100 × 20 mm, TC treated, a bag of 10 pieces, (# 0030702.115)
- Corning Cell Culture T25 flasks, TC treated, (#430168)
- 1 mL syringe (DISPO van syringe)
- GEN-CT-15ML-S15 m, Centrifuge Tube with Cap, Conical, Clear, Sterile- 25 × 20, Axygen (#39172200)

[a] Although the catalog number and supplier are mentioned for reference, comparable products can be bought at a comparable price from several sources.

- GEN-CT-50ML-S, 50 mL Centrifuge Tube with Cap, Conical, Clear, Pre-Sterilized, Axygen (#39172200)

2.2 Cells and reagents

- *M. smegmatis* [mc2 155] (and recombinant strains thereof)
- THP-1 cell line (#TIB-202) ATCC
- HEK293T cell line (CRL-3216 ™) ATCC
- RPMI-1640 (#11875119, GIBCO)
- Antibodies (appropriate primary and secondary)
- Mitochondrial resuspension buffer [10 mM Tris-Cl, pH 7.6]
- Trypan blue, 0.4% (Thermo Fischer, # 15250061),
- Bradford reagent (Thermo Fischer, # 232000),
- Electrophoresis reagents (Qualigens, India)
- Lipofectamine ™ 2000 (Invitrogen™, # 1668027)
- Dulbecco's Modification of Eagle's Medium (DMEM), 4.5 g/L glucose (#15–017-CV, GIBCO)
- Fetal Bovine Serum (FBS) (# 10270106, GIBCO)
- Pen-strep (100×) (HiMedia, # A001A)
- Phosphate-buffered saline (PBS) (#10010023, Life Technologies)
- Trypsin-EDTA (0.5%), no phenol red (#15400054, Thermo Fisher Scientific)
- 7H9 (HIMEDIA, India #M198)

2.3 Equipment

- Cell culture facility (BSL2) with phase contrast microscope (Leica with 10× and 40× Objective lenses),
- Centrifuge with swingout rotor (Eppendorf, Germany),
- Cell culture incubator (Eppendorf, Germany),
- Micropipettes (Eppendorf, Germany),
- 2 mL Dounce homogenizer (Borosil),
- Mini vertical electrophoresis unit (Bio-rad, Germany),
- Western transfer unit (Bio-Rad, Germany),
- Rocker (Neuation, India)
- Benchtop spectrophotometer for nucleic acid quantification (Eppendorf, Germany),
- Spectrophotometer (Thermo Fisher Scientific, USA)
- Eppendorf® PCR Cooler (#Z606634, Millipore-Sigma)
- Humidified cell culture incubator (Eppendorf, Germany)

- Laboratory biosafety cabinet (Class II), Class II Type A2 Biological Safety Cabinet

2.4 Software

- Image Lab
- Image J
- Las X software
- Huygens software

3. Methods

This chapter is instructive to those researchers, who need to explore a given pathogen for possible mitochondria-targeting proteins or to test if a given protein targets host mitochondria. Several websites such as MITOPROT, and PSORT II, offer to analyze if a given protein sequence (with amino acid sequence input) has a mitochondrial targeting potential. The output of the analysis returns the possibility of the protein entering host mitochondria with a confidence score. Taking examples of both positive and negative control proteins, which are known to either target host mitochondria or not, will help in setting up high confidence scores.

The following protocol has been described to investigate if a given *Mtb* protein that is predicted to target mitochondria is indeed targeted by using both biochemical and confocal approaches. Towards this, the candidate genes with good confidence scores, MTP-1 and MTP-2 were cloned in both mycobacterial and mammalian vector systems.

3.1 Expression of a candidate gene in mycobacterial vector

- The gene is cloned[b] in mycobacterial vector,[c] pMSP::mCherry[d] (Addgene, Plasmid #30169) with a Cherry tag.
- The sequence-verified construct is amplified and used for transformation of *M. smegmatis* (Section 3.3).

[b] MTP-1 and MTP-2 were cloned in *Hin*dIII and *Nde*I restriction enzyme sites.

[c] A mycobacterial vector is a shuttle vector that could be amplified in an *E. coli* system by virtue of an *E. coli* origin of replication and potentially expresses in a mycobacterial system by virtue of mycobacterial strong promoter, hsp60.

[d] mCherry provides a fluorescent tag with Emi/Ex wavelength of 587 nm/610 nm.

3.2 Expression of the candidate gene in mammalian vector

- The gene of interest is cloned in pEGFPN1[e] (or any fluorescent/tag[f]).
- The sequence-verified construct is amplified and used for transfection (Section 3.6). The expression of the protein of interest is confirmed by immunoblotting with an anti-tag antibody or anti-sera raised against the specific protein.

3.3 Transformation of *Mycolicibacterium smegmatis*

- The sequence-verified plasmid[g] construct (1 μg) is incubated with *M. smegmatis* competent cells[h] for 10 min on ice and electroporated using an electroporator (Eppendorf) for 5 ms at 2500 mV using pre-chilled electroporator cuvettes.
- To the electroporated *M. smegmatis* cells, 5 mL of 7H9 media is added and incubated at 37 °C for 2 h.
- After incubation, the cells are harvested by centrifugation at 5000 × *g* for 5 min and plated on 7H10 plates containing 34 μg/mL kanamycin.
- One of the colonies[i] that appears on the plate is inoculated and grown in 7H9 media containing 2% glycerol and 34 μg/mL kanamycin.
- A secondary culture is grown until OD_{600nm} of 1.0 and stored in 30% glycerol stocks.[j]

3.4 Culture of host cells

- Human THP-1[k] (monocyte) cells are grown in RPMI-1640 complete medium (HIMEDIA, India) and HEK293T cells[l] in DMEM (Cat no

---

[e] MPT-1 and MPT-2 are cloned in the *Xho*I/BamH1 sites of pEGFPN1 or *Hin*dIII/*Eco*RI sites of pCDNA3.1.

[f] In this study, we cloned MPT-1 in pcDNA3.1 myc-His. Alternatively, one could use pEGFPN1, pcDNA3-YFP or pLVX-mcherry-C1.

[g] The plasmid should be devoid of excess salt as it would hinder electroporation.

[h] *M. smegmatis* competent cells are prepared following the method outlined by Tanya Parish. Briefly, *M. smegmatis* grown at OD600nm of 1.0 are harvested and cooled to 0 °C on ice and resuspended in ice cold-10% glycerol and washed twice with ice-cold- 10% glycerol with reducing the culture volume to half every time. Final resuspension is done in a small volume (400 μL) and stored in cryovials at −80 °C.

[i] The dense cultures expressing the cherry tag are visualized as pink culture.

[j] For glycerol stock preparation, 10 mL of the freshly growing culture is harvested at 5000 × *g* for 5 min and resuspended in 5 mL fresh medium containing 30% glycerol.

[k] THP-1 cells are human monocyte suspension cells of Acute monocytic leukemia origin isolated from peripheral blood. They are spherical in the monocytic stage, but readily form appendages upon activation to macrophages and adhere strongly to treated tissue culture plates.

[l] HEK293T cells are adherent cells and need mild trypsin (0.25×) treatment for harvesting and subculturing.

11965092, Invitrogen) with 10% Fetal Bovine Serum (FBS) (Gibco, USA) and 1× Antibiotic-antimycotic (GIBCO, USA) at 37 °C in 5% CO_2.
- The culture is monitored regularly and replenished with fresh media as required.

3.5 Mycobacterial culture and infection

- *M. smegmatis* is cultured in 7H9 medium (HIMEDIA, India) supplemented with 2% glycerol.
- Before infection, the mycobacterial culture is stained using a Zeil Neelson staining kit to rule out any possible contamination before use for experiments.
- The contamination-free mycobacterial culture is used for the infection of THP-1 cells.
- The mycobacterial suspension is passed through a 1 mL syringe several times (minimum of 5–6 times) to remove any aggregates that affect the infectivity.
- OD_{600nm} is recorded and the number of bacilli is calculated using the McFarland Nephelometer standards formula,

$$\text{No of bacteria per mL} = 1.5 \times 10^8 \text{ x } OD_{600nm}/0.132$$

- For infection, around 1 million THP-1 cells are seeded in each well of a 6-well plate and differentiated into macrophages with PMA (phorbol-12-myristate-13-acetate) at a final concentration of 25–100 ng/mL and incubated overnight.
- Subsequently, the media is replaced with fresh pre-warmed RPMI media for 24 h as a rest period.
- Further, the THP-1 macrophages are infected with 5–10 million wild-type *M. smegmatis* (or its recombinant derivative overexpressing a candidate gene, whose role in mitochondrial interaction is sought for) (MOI of 20) for around 4 h in antibiotic-free media and replaced with fresh pre-warmed RPMI media. The infected cells are incubated for another 4 h in the 37 °C incubator.

3.6 Transfection

- HEK293T cells[m] are used to transfect either pEGFPN1 or pEGFPN1-MPT1/2 constructs using lipofectamine 2000 (Invitrogen, USA).

[m] HEK293T cells offer great transfection efficiency.

- Briefly, around 0.5 million are seeded in each well of a 6-well plate and incubated overnight.
- The following day, the media is replaced with 1.8 mL fresh pre-warmed media.
- The DNA-lipofectamine complex is made by mixing 1 μg DNA of the respective plasmids (diluted to 100 μL with Opti-MEM) and 1 μL of lipofectamine (diluted to 100 μL with Opti-MEM) and incubated at 37 °C for 10 min.
- The complex is added to the cells in the 6-well plate drop-wise covering the entire well and mixed by gentle swirling.
- Further, the cells are incubated for 6 h at 37 °C in a 5% CO_2 incubator. The media is replaced with fresh pre-warmed DMEM and incubated for 24–48 h.

3.7 Imaging of mycobacterial infection

- The infected live cells are washed thrice with PBS and stained with Hoechst dye (Thermo Fisher Scientific, USA Cat: H3570) for 20 min at 37 °C in 5% CO_2.
- The stained cells are washed thrice with PBS and replaced with fresh pre-warmed RPMI media.
- The images are captured using a Carl Zeiss confocal laser scanning microscope under 37 °C with 5% CO_2.
- A video is collected for 30 min with images collected every 10–30 s. Snapshots of the cells under different fluorescence channels are visualized for mycobacterial infection and secretion of cherry-tagged protein (MPT-3) (Fig. 1).

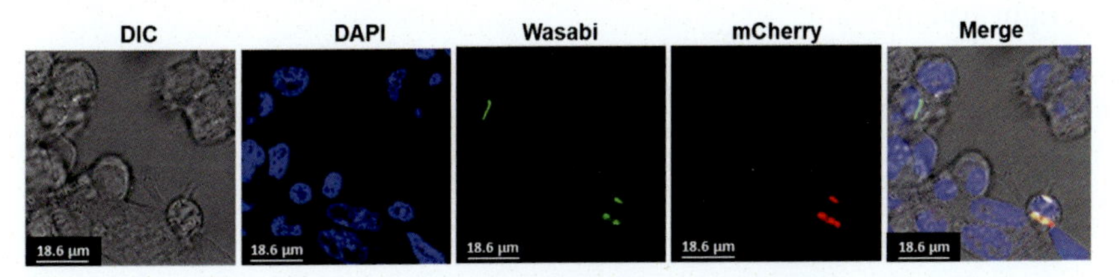

Fig. 1 Infection of THP-1 cells with fluorescently stained *Mycolicibacterium smegmatis*. THP-1 macrophages were infected with *M. smegmatis* (Wasabi/mCherry) at an MOI of 20 as described in Section 3.5 and imaged by live cell analyses as described in Section 3.7. The internalized *M. smegmatis* (mCherry-tagged proteins could be used in combination with appropriate counter mitochondrial stains to visualize the mitochondrial localization of cherry-tagged proteins PBS with 0.1% Tween 20) expressing mCherry (red)/ Wasabi (green) can be visualized as yellow spots. The white line in the panel indicates a scale bar of 18.6 μm.

3.8 Fluorescence staining and imaging

- HEK293T cells are transfected with mitoGFP (1 μg) as well as pCDNA-myc or pCDNA-MPT1 on a cover slip placed in a 6-well plate.
- Post transfection, the cells are washed thrice with PBS and fixed in 4% paraformaldehyde (SIGMA, USA) for 15–20 min at room temperature (~22-25 °C).
- The fixed cells are washed again with PBS thrice and permeabilized with 0.5% Triton-X-100 (Qualigens, India) for 15 min followed by washing the cells twice with PBS.
- The permeabilized cells are blocked using 3% BSA (SRL, India) for 30 min.
- The cells are washed once with PBS and incubated in an anti-myc antibody (in 1% BSA at 1:2000 dilution) for 2 h, followed by three washes with PBS.
- The cells are further incubated with the corresponding secondary antibody tagged with Texas Red (1:5000 dilution) for 1 h.
- Subsequently, the cells are washed with PBS and the nuclei are stained with DAPI (Abcam, USA).
- Imaging is done by Leica confocal microscopy and visualized by Las X software. Plot profiles are generated by using Image J and Huygens software. Images of around 5–10 fields are captured randomly per slide by Leica imaging Software to visualize GFP (Ex/Emi: 488/510 nm) as green fluorescence and over-expressed protein as Texas Red (Ex/Emi: 596/615 nm) as red fluorescence.
- The merged images are visualized, wherein yellow fluorescence reflects the co-localization of GFP and mitochondria. The individual green and red fluorescence upon merging indicates no colocalization (Fig. 2, right panel).

Generation of plot profile

- Convert the image file to PNG format and open it in Image J (Fiji) software
- A line is drawn across where the mitochondrial localization is observed (as shown in inset, Fig. 2).
- The line is selected as the region of interest (ROI) under 'Analyze' tool in Image J (Analyze Tools Region of Interest)
- Apply the 'region of interest' to the individual images showing the protein signal and mitochondrial signal
- Go to Analyze, and select 'Plot Profile' for individual images.
- Select 'List' in the plot Profile Menu

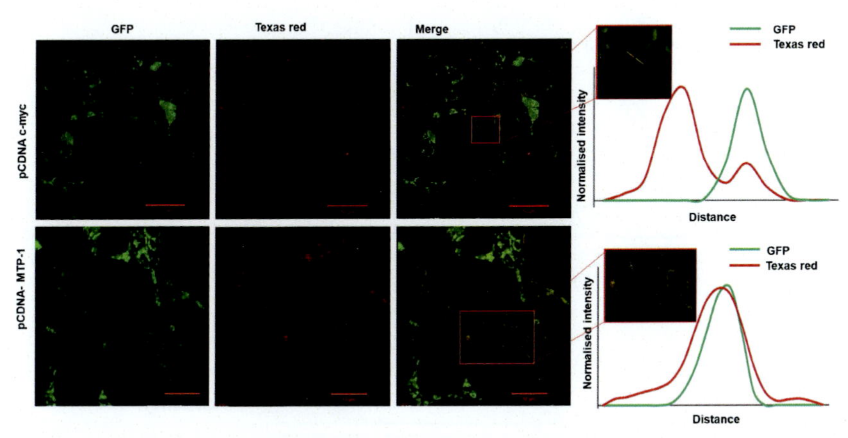

Fig. 2 Immunofluorescence analysis of MPT-1 expressing host cells. HEK293T cells were transfected with either pCDNA-c-myc or the expression construct pCDNA-MPT-1 together with mito-GFP (stains mitochondria green). 24h Post-transfection, the cells were processed for immunofluorescence as described in Section 3.8 for myc tag by Texas Red (represents myc tagged protein) and imaged by confocal microscopy. Plot profiles of the mitochondrial and protein (myc-tag) localization in the cell were plotted to confirm if the protein localizes to the mitochondria as also seen in the confocal merged picture as yellow stained loci. The inset shows the line that was used for constructing the plot profile as described in Section 3.8. The red line in pCDNA c-myc panel indicates a scale bar of 50 μm and in pCDNA MTP-1 panel indicates 10 μm.

- Copy the values into Excel and normalize with max intensity for both mitochondria and protein signals, generate the graph, and smoothen the curves (Fig. 2).

3.9 Isolation of mitochondria

- The infected THP-1 macrophages are treated with 0.25 X trypsin (1 mL per 100 mm dish) for 5 min at 37 °C and harvested by gently scraping in 1× PBS and harvested at 1000 × *g* for 5 min and repeated twice.
- Further, the cells are harvested and resuspended in 1 mL of mitochondrial re-suspension buffer (10 mM Tris, pH 7.6 with 1× PIC), and lysed with a 2 mL Dounce homogenizer.
- Around 200 μL of 1.5 M sucrose solution is added to the homogenate, mixed, and centrifugated at 600 × *g* for 10 min at 2 °C.
- The supernatant containing the cell lysate is collected and again centrifuged at 14,000 × *g* for 10 min at 2 °C.
- The supernatant containing the cytosol is processed for acetone precipitation. Towards this, four volumes of acetone are added to one volume of cytosolic fraction, mixed, and incubated at −20 °C overnight.

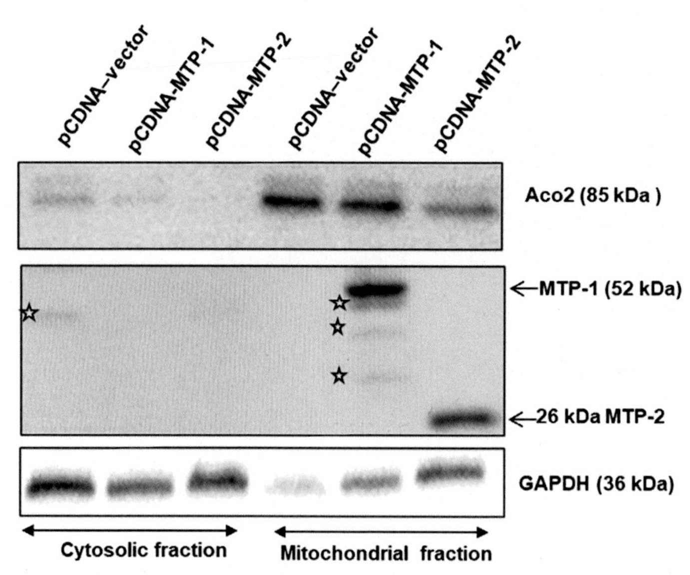

Fig. 3 Isolation and enrichment of mitochondria for confirmation of mitochondrial targeting proteins. HEK293T cells were transfected with either pCDNA3.1 or pCDNA-MTP-1 or pCDNA-MTP-2 as described in Section 3.6 and 24 h later, the cells were harvested and processed for mitochondrial enrichment as detailed in Section 3.9. Both the cytoplasmic and the enriched mitochondrial proteins were fractionated on 12% SDS-PAGE and analyzed for the presence of MTP-1 (52 kDa) and MTP-2 (26 kDa) in the host mitochondria as outlined in Section 3.10. Aconitase (85 kDa) was used as a marker for mitochondria and GAPDH (36 kDa) as a cytoplasmic marker. Please note that the mitochondrial enrichment is not 100% pure mitochondria preparation and has residual content of cytoplasm (as reflected by the presence of GAPDH in mitochondrial fraction). Stars in the figure indicate non-specific proteins.

- The precipitate is collected by centrifugation at 14,000 × *g* for 10 min and the supernatant is discarded.
- The precipitate is resuspended in 40 μL of ice-cold mitochondrial re-suspension buffer and an equal concentration of mitochondrial proteins as quantified by Bradford (Thermo Fischer, USA) from different experimental samples are used for immunoblotting analyses (Fig. 3).

3.10 Immunoblotting analyses

- The protein concentration in all samples is estimated by Bradford method
- An equal amount (50 μg) of protein from each sample is fractionated on 12% SDS-PAGE

- Two sets of the fractionated proteins are transferred onto nitrocellulose membrane by using standard Towbin buffer and semi-dry apparatus, 30 min, 15 V (Biorad USA)
- The membranes are blocked using 5% non-fat dry milk powder (Himedia, India, # TC791) in 1× PBS.
- The membranes are washed twice with PBST and incubated with either anti-myc antibody in one set and the other membrane is cut into two halves and probed with anti-Aco2 antibody (1: 2000 dilution) (as a marker for mitochondrial fraction, 85 kDa) with the upper half and probed with anti-GAPDH (1:5000) (as a marker for cytoplasmic fraction, 36 kDa) using the lower half of membrane overnight at 4 °C
- All the membranes are washed 4 times with PBST for 5 min.
- The membranes are incubated with appropriate secondary antibody (@1: 10000 dilution) [goat- anti-mouse for myc protein and mouse anti-rabbit for Aco-2 and. Rabbit-anti-mouse for GAPDH] at room temperature for 2 h.
- All the membranes are washed 4 times with PBST for 5 min and finally incubated in PBS until development.
- The membranes are developed using Advansta WesternBright™ ECL detection kit (#K-12045) in a chemidoc (Bio-Rad, USA) and the images are captured.

4. Concluding remarks

Mitochondria are the most dynamic organelles of a cell and are the obvious target of choice for several intracellular pathogens as well as the seat for metabolic and immunological pathologies including cardiovascular diseases, cancer, diabetes, inflammation, obesity, and aging. Therefore, they are also the site of pharmacological correction and offer the platform for host-directed mitochondria therapy (Li & Huang, 2020). Dysfunctional mitochondria can be corrected by mitochondrial transplantation, targeting small molecules, activation of mitophagy, and blocking of mitochondrial fission and fusion (Videla, Marimán, Ramos, José Silva, & del Campo, 2022; Zong et al., 2024).

Acknowledgments

The work described in the chapter was funded by a project sanctioned to SB by SERB SUPRA (SPR/2021/000137). pMSP12::mCherry was a gift from Lalita Ramakrishnan (Addgene, plasmid #30169; http://n2t.net/addgene:30169; RRID:Addgene_30169). The mito-GFP construct was kindly gifted by Prof. Naresh Babu Sepuri, University of Hyderabad.

Conflicts of interests

The authors declare no conflict of interest.

References

Asalla, S., Mohareer, K., & Banerjee, S. (2017). Small molecule mediated restoration of mitochondrial function augments anti-mycobacterial activity of human macrophages subjected to cholesterol induced asymptomatic dyslipidemia. *Frontiers in Cellular and Infection Microbiology*, 7, 439. https://doi.org/10.3389/fcimb.2017.00439. PMID: 29067283; PMCID: PMC5641336.

Behl, T., Makkar, R., Anwer, M. K., Hassani, R., Khuwaja, G., Khalid, A., et al. (2023). Mitochondrial dysfunction: A cellular and molecular hub in pathology of metabolic diseases and infection. *Journal of Clinical Medicine*, *12*. https://doi.org/10.3390/jcm12082882.

Brokatzky, D., & Häcker, G. (2022). Mitochondria: Intracellular sentinels of infections. *Medical Microbiology and Immunology*, *211*, 161–172. https://doi.org/10.1007/s00430-022-00742-9.

Cadieux, N., Parra, M., Cohen, H., Maric, D., Morris, S. L., & Brennan, M. J. (2011). Induction of cell death after localization to the host cell mitochondria by the mycobacterium tuberculosis PE-PGRS33 protein. *Microbiology*, *157*, 793–804. https://doi.org/10.1099/mic.0.041996-0.

Cartes-Velásquez, R., Vera, A., Antilef, B., Sanhueza, S., Lamperti, L., González-Ortiz, M., et al. (2024). Metformin restrains the proliferation of CD4+ T lymphocytes by inducing cell cycle arrest in Normo- and hyperglycemic conditions. *Biomolecules*, *14*. https://doi.org/10.3390/biom14070846.

Dubey, R. K., Dhamija, E., Kumar Mishra, A., Soam, D., Mohanrao Yabaji, S., Srivastava, K., et al. (2021). Mycobacterial origin protein Rv0674 localizes into mitochondria, interacts with D-loop and regulates OXPHOS for intracellular persistence of mycobacterium tuberculosis. *Mitochondrion*, *57*, 241–256. https://doi.org/10.1016/j.mito.2020.11.014.

Elesela, S., & Lukacs, N. W. (2021). Role of mitochondria in viral infections. *Life*, *11*. https://doi.org/10.3390/life11030232.

Gao, K., Cheng, M., Zuo, X., Lin, J., Hoogewijs, K., Murphy, M. P., et al. (2021). Active RNA interference in mitochondria. *Cell Research*, *31*, 219–228. https://doi.org/10.1038/s41422-020-00394-5.

Jamwal, S., Midha, M. K., Verma, H. N., Basu, A., Rao, K. V. S. S., & Manivel, V. (2013). Characterizing virulence-specific perturbations in the mitochondrial function of macrophages infected with mycobacterium tuberculosis. *Scientific Reports*, *3*, 1328. https://doi.org/10.1038/srep01328.

Joseph, S., Yuen, A., Singh, V., & Hmama, Z. (2017). Mycobacterium tuberculosis Cpn60.2 (GroEL2) blocks macrophage apoptosis via interaction with mitochondrial mortalin. *Biology Open*. https://doi.org/10.1242/bio.023119.

Kumar, R., Singh, P., Kolloli, A., Shi, L., Bushkin, Y., Tyagi, S., et al. (2019). Immunometabolism of phagocytes during *Mycobacterium tuberculosis* infection. *Front. Mol. Biosci*. https://doi.org/10.3389/fmolb.2019.00105.

Li, Q., & Huang, Y. (2020). Mitochondrial targeted strategies and theirapplication for cancer and other diseases treatment. *Journal of Pharmaceutical Investigation*, *50*, 271–293. https://doi.org/10.1007/s40005-020-00481-0.

Lum, M., & Morona, R. (2014). Dynamin-related protein Drp1 and mitochondria are important for Shigella flexneri infection. *International Journal of Medical Microbiology*, *304*, 530–541. https://doi.org/10.1016/j.ijmm.2014.03.006.

Marchi, S., Morroni, G., Pinton, P., & Galluzzi, L. (2022). Control of host mitochondria by bacterial pathogens. *Trends in Microbiology, 30*, 452–465. https://doi.org/10.1016/j.tim.2021.09.010.

Medikonda, J., Wankar, N., Asalla, S., Raja, S. O., Yandrapally, S., Jindal, H., et al. (2024). Rv0547c, a functional oxidoreductase, supports mycobacterium tuberculosis persistence by reprogramming host mitochondrial fatty acid metabolism. *Mitochondrion, 78*, 101931. https://doi.org/10.1016/j.mito.2024.101931.

Mohareer, K., Asalla, S., & Banerjee, S. (2018). Cell death at the cross roads of host-pathogen interaction in mycobacterium tuberculosis infection. *Tuberculosis*. https://doi.org/10.1016/j.tube.2018.09.007.

Mohareer, K., & Banerjee, S. (2023). Mycobacterial infection alters host mitochondrial activity. *Academic Press.*. https://doi.org/10.1016/bs.ircmb.2023.01.007.

Mohareer, K., Medikonda, J., Vadankula, G. R. G. R., & Banerjee, S. (2020). Mycobacterial control of host mitochondria: Bioenergetic and metabolic changes shaping cell fate and infection outcome. *Frontiers in Cellular and Infection Microbiology, 10*. https://doi.org/10.3389/fcimb.2020.00457.

Rana, A., Rub, A., & Akhter, Y. (2014). Proteome-scale identification of outer membrane proteins in Mycobacterium avium subspecies paratuberculosis using a structure based combined hierarchical approach. *Molecular BioSystems, 10*, 2329–2337. https://doi.org/10.1039/c4mb00234b.

Sohn, H., Kim, J.-S. S., Shin, S. J., Kim, K., Won, C.-J. J., Kim, W. S., et al. (2011). Targeting of mycobacterium tuberculosis heparin-binding hemagglutinin to mitochondria in macrophages. *PLoS Pathogens*, 7, e1002435. https://doi.org/10.1371/journal.ppat.1002435.

Tiku, V., Tan, M. W., & Dikic, I. (2020). Mitochondrial functions in infection and immunity. *Trends in Cell Biology, 30*, 263–275. https://doi.org/10.1016/j.tcb.2020.01.006.

Tundup, S., Mohareer, K., & Hasnain, S. E. (2014). Mycobacterium tuberculosis PE25/PPE41 protein complex induces necrosis in macrophages: Role in virulence and disease reactivation? *FEBS Open Bio, 4*, 822–828. https://doi.org/10.1016/j.fob.2014.09.001.

Videla, L. A., Marimán, A., Ramos, B., José Silva, M., & del Campo, A. (2022). Standpoints in mitochondrial dysfunction: Underlying mechanisms in search of therapeutic strategies. *Mitochondrion, 63*, 9–22. https://doi.org/10.1016/j.mito.2021.12.006.

Zong, Y., Li, H., Liao, P., Chen, L., Pan, Y., Zheng, Y., et al. (2024). Mitochondrial dysfunction: Mechanisms and advances in therapy. *Signal Transduction and Targeted Therapy, 9*, 124. https://doi.org/10.1038/s41392-024-01839-8.

CHAPTER TEN

Bacterial predators and BALOs: Growth protocol and relation with mitochondria

Valerio Iebba*
Gustave Roussy Cancer Campus, Villejuif, France
*Corresponding author: e-mail address: viebba@units.it

Contents

Abstract

The microbial world is characterized by mechanisms of competition and predation, akin to the animal world. However, while predation's ecological role is well-established in animals, it's less understood in bacteria due to fewer known predators and unclear phylogenetic affiliations. Nevertheless, microorganisms can prey on bacterial cells, including Bacteriophages, Protists, and Predatory Prokaryotes. These predators inhabit various habitats and may play vital roles in bacterial ecology and ecosystem regulation. Predatory interactions between host and parasite are common in nature. Predatory bacteria, such as *Bdellovibrio* and like organisms (BALOs), employ various strategies, including epibiotic predation and direct invasion. BALOs, which thrive in the periplasmic space of Gram-negative bacterial cells, modulate bacterial populations and could serve

Methods in Cell Biology, Volume 194
ISSN 0091-679X
https://doi.org/10.1016/bs.mcb.2024.07.003

as preventive or therapeutic agents against Gram-negative infections. While primarily active against extracellular prey, BALOs may also target mitochondria, which are crucial for cellular processes. The relationship between intracellular bacteria and host mitochondria, including morphology, function, and apoptosis, warrants further exploration. Protocols for growing, propagating, and detecting predatory activities of BALOs, particularly *Bdellovibrio bacteriovorus*, are provided to assess their presence and activities against potential prey.

1. Introduction

The microbial world is characterized by mechanisms of competition and predation, similar to the animal world. However, in the latter, the ecological role of predation is well established and dominates every trophic level. This is not the case in the bacterial realm, where the number of known bacterial predators is smaller, and their phylogenetic affiliations are widely unknown. Nevertheless, at the microbial level, there are microorganisms capable of preying on bacterial cells. Bacterial predators include Bacteriophages, Protists, and Predatory Prokaryotes (Balows et al., 2013; Casida, 1988; Cohen et al., 2021; Jurkevitch & Mitchell (Professor), 2020; Lambert et al., 2003; Schwudke et al., 2001; Sockett, 2009). These predators can be found in various habitats, such as rivers, saltwater, freshwaters, sewage, soils, plant roots, and animal feces. They may play a crucial role in the bacterial ecology of natural environments, the ecological balance, and the regulation of bacterial populations in natural ecosystems (Iebba et al., 2013; Jurkevitch & Mitchell, 2020; Kessel & Shilo, 1976; Lambina et al., 1987; Markelova, 2010; Pérez et al., 2016; Shilo & Bruff, 1965; Singh et al., 2022; Sockett, 2009, 2023). Regardless of the mechanism of predation, the predatory interactions between the "host" and "parasite" in nature seem to be the rule rather than the exception (Casida, 1988; Fallon & Brock, 1979; Guerrero et al., 1986; Jurkevitch & Mitchell, 2020; Martin, 2002). The currently best-known example of a predatory bacterium is *Bdellovibrio*, but other predatory bacteria have been collectively described under the name of *Bdellovibrio* and like organisms (BALOs) (Balows et al., 2013; Casida, 1988; Jurkevitch, 2006).

Most predatory bacteria are extracellular, such as the genera *Ensifer*, *Vampirovibrio*, *Vampirococcus*, *Micavibrio*: the predator attaches to the outer surface of the host cell, beginning to degrade and assimilate host molecules through specialized structures. This type of bacteria employs a typical predatory strategy called epibiotic (Esteve et al., 1983; Esteve & Gaju, 1999; Guerrero et al., 1986; Jurkevitch & Mitchell, 2020). Moreover, other bacteria such as myxobacteria, *Cytophaga*, or *Herpetosiphon* are capable of lysing and utilizing living bacterial cells as a food substrate. In fact, these bacteria commonly produce a variety of hydrolytic enzymes that degrade prey cells, leading to the localized availability of host cell-derived nutrients. This predatory approach is called the "wolfpack" strategy or "group predation." Examples of this kind of predation are *Myxococcus* (Burnham et al., 1968; Fallon & Brock, 1979) and *Lysobacter* (Lin & McBride, 1996), and, at a lesser extent, for *Bdellovibrio bacteriovorus* (Iebba et al., 2014; Jurkevitch & Mitchell, 2020; Pantanella et al., 2018). A third approach to predation, called direct invasion or endobiotic predatory strategy, occurs when a predator invades the prey cytoplasm, as seen in *Daptobacter* (Cohen et al., 2021; Johnke et al., 2020; Martin, 2002). The best-characterized examples of predatory bacteria belong to the order of Bdellovibrionales and Bacteriovoracales. They survive by preying on other bacteria, particularly acting as obligate predators of Gram-negative bacterial cells (Iebba et al., 2014; Jurkevitch & Mitchell, 2020; Jurkevitch & Mitchell (Professor), 2020; Pantanella et al., 2018; Sockett, 2023). They invade, grow, and replicate within a specific compartment found in Gram-negative cells known as the periplasmic space. Inside this space, they undergo a complex developmental cycle that culminates in the killing of the prey cell and the release of progeny cells outside (Caulton et al., 2024; Jurkevitch, 2006; Lambert et al., 2003; Núñez et al., 2003; Sockett, 2023; Stolp & Starr, 1963) (Fig. 1). BALOs can rapidly reduce bacterial populations, including pathogens, through their lytic action; for this reason, they are considered as biological modulators of the type and density of bacterial populations in nature, as well as potential preventive or therapeutic agents against infections by Gram-negative bacteria (Horowitz et al., 1974; Lambert et al., 2009; Sockett, 2009, 2023).

Even if active against extracellular bacterial prey, BALOs could be potentially active against mitochondria, if accessed to them. Mitochondrial function is closely linked to cellular processes like energy production, immunity, and apoptosis, making them attractive targets for pathogens and, potentially, their BALOs predators. The relationship between intracellular

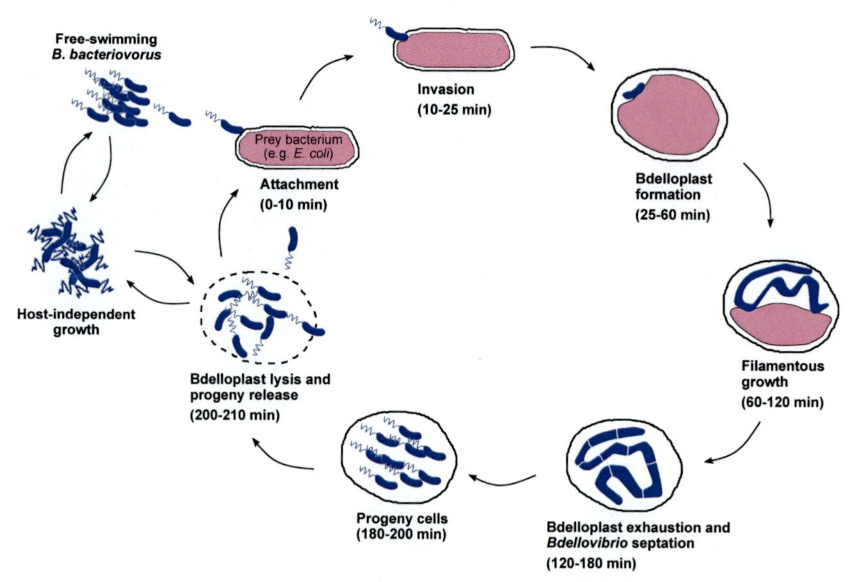

Fig. 1 Life cycle of *Bdellovibrio bacteriovorus*. The predator could survive indefinitely without prey in a host-independent manner (HI), or going through a complex prey intra-periplasmic cycle encompassing the formation of a bdelloplast and progeny release (total duration of around 210 min).

bacteria and host cell mitochondria, particularly regarding morphology, function, and apoptosis, is well explained in recent papers (Marchi et al., 2022; Maurice & Sadikot, 2023; Sacchi et al., 2004; Spier, Stavru, & Cossart, 2019), but a few pointed out a sight on intra-mitochondrial bacteria (Comandatore et al., 2021; Stavru et al., 2020; Uzum et al., 2023), on BALOs as the phylogenetic origin of mitochondria (Bremer, Tria, Skejo, Garg, & Martin, 2022; Davidov & Jurkevitch, 2009; Kamada, Wakabayashi, & Naganuma, 2023), or on BALOs action on eukaryotic cells (Gupta et al., 2016), thus leaving space for future developments.

Regarding the most prominent species among BALOs, *B. bacteriovorus*, here we describe the protocols (Fig. 2) on how to: (i) grow and propagate it; (ii) assess its predatory activities; (iii) detect it by means of species-specific end-point and qPCR. These three protocols, even applicable to other BALOs with little modifications, are useful to assess the presence and predatory activities of BALOs against potential prey.

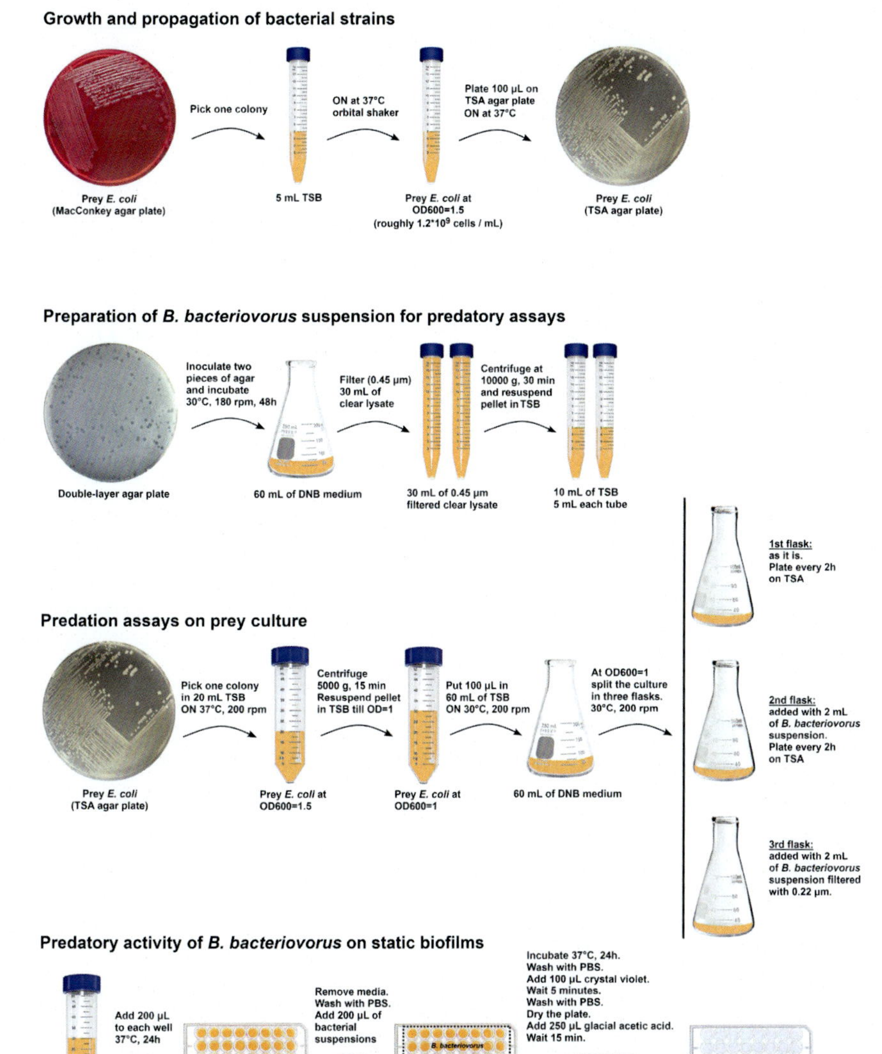

Fig. 2 Workflow visualization of the protocols. The figure shows the protocols 3.1 (Growth and propagation of bacterial strains), 3.2 (Preparation of *B. bacteriovorus* suspension for predatory assays), 3.3 (Predation assays on prey cultures) and 3.4 (Predatory activity of *B. bacteriovorus* on static biofilms). The protocol 3.5 on qPCR is not reported.

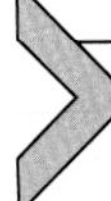

2. Materials

2.1 Common disposables

- 10, 20, 200 and 1000 μL Sartorius filter pipette tips (#Z757764, #Z757772, #Z757799, #Z757810, Merck) (see **Notes 1, 2**)
- 15 mL conical tube (#CLS430766, Merck) (see **Notes 1, 2**)
- 50 mL conical tube (#CLS431472, Merck) (see **Notes 1, 2**)
- UVette Eppendorf (#Z618683, Merck) (see **Note 1**)
- Inoculating loop volume 1 μL (#HS81121A, Merck) (see **Note 1**)
- 48-well plate (#CLS356509, Merck) (see **Notes 1, 2**)
- Syringe filter unit 0.45 μm pore size (#WHA99111304, Merck) (see **Note 1**)
- L-shape spreader (#HS8171B, Merck) (see **Note 1**)
- Syringe 20 mL w/o needle (#Z116882, Merck) (see **Note 1**)
- Hard-Shell® 96-Well PCR Plates, low profile, thin wall, skirted, white/clear (#HSP9601, Bio-Rad) (see **Note 1**)
- Microseal "B" PCR Plate Sealing Film, adhesive, optical (#MSB1001, Bio-Rad)

2.2 Cells and reagents

- *B. bacteriovorus* strain HD100 (#50701, DSMZ) (see **Note 3**)
- *Pseudomonas aeruginosa* strain PAO (#1707, DSMZ) (see **Note 4**)
- DNB growth medium contained: 0.8 g/L Bacto Nutrient Broth (NB) and 0.1 g/L yeast extract, with the separated addition of 0.45 μm-filtered 0.3 g/L of Casamino acids, 0.5 g/L of $CaCl_2{\cdot}2H_2O$ and 0.6 g/L of $MgCl_2{\cdot}6H_2O$ (see **Note 5**)
- Taq DNA-polymerase (#D9307, Merck) (see **Note 1**)
- Bovine serum albumin (BSA) (#810533, Merck) (see **Note 1**)
- GelRed 10,000× (#STC123, Merck)
- 2× qPCR ProbesMaster (#PCR-360S, Jena Bioscience GmbH) (see **Note 1**)
- Primers for PCR and qPCR, probe for qPCR (Eurofins GmbH, Ebersberg, Germany) (see **Note 1**)
- Plasmid (pUC57) containing one copy of the 16S rDNA target (Eurofins GmbH, Ebersberg, Germany) (see **Note 1**)
- Tryptone Soy Agar TSA (#1.46004, Merck) (see **Note 1**)
- Tryptone Soy Broth TSB (#STBMTSB12, Merck) (see **Note 1**)

- Phosphate-buffered saline (PBS) (#10010023, Life Technologies) (see **Note 1**)
- Crystal Violet Solution 1% (#V5265, Merck) (see **Note 1**)
- Glacial acetic acid (#695092, Merck) (see **Note 1**)
- Agarose (#A9539, Merck) (see **Note 1**)
- PCR-grade distilled water (#PCRH20-RO, Merck) (see **Note 1**)

2.3 Equipment (see Note 1)

- Eppendorf BioPhotometer™ D30 (#6133 000.001, Eppendorf)
- Incubator/shaker (#5000IR, VWR)
- Savant™ SpeedVac™ (#15819206, Thermo Fisher Scientific)
- microDOC Compact Gel Documentation System, with UV Transilluminator (UVT365) (#Z692743, Merck)
- NanoDrop Eight Spectrophotometer (#NDE-GL, Thermo Fisher Scientific)
- Eppendorf benchtop centrifuge 5810G (#EP5810000420, Merck)
- Binder drying oven (thermostat) E28 series (#Z603929, Merck)
- DM4 B Upright Microscopes (Leica Microsystems)
- Benchmark SmartReader™ 96 Microplate Absorbance Reader with Incubation (#BMSMR9600TE, Merck)
- Eppendorf® Mastercycler® Thermocycler (#EP6313000018, Merck)
- CFX Opus 96 Real-Time PCR System (#12011319, Bio-Rad)

2.4 Softwares (see Note 1)

- TotalLab TL120 software (Non-linear Dynamics)
- Prism 5 software (GraphPad, La Jolla, California, USA)
- Phoretix 1D software (TotalLab, Newcastle upon Tyne, United Kingdom)

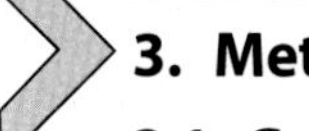

3. Methods

3.1 Growth and propagation of bacterial strains

1. Prey strain (one colony from a Petri dish) are first suspended in 5 mL of Tryptone Soya Broth (TSB) put into a 15 mL conical tube, and let to grow overnight at 37 °C in an orbital shaker incubator
2. To verify the purity of the culture, 100 μL of overnight culture are spread onto Tryptone Soya Agar (TSA) plates by means of L-shaped spreaders (see **Note 6**)

3. Cultivate the predator strain *B. bacteriovorus* as previously reported (Jurkevitch, 2006; Kadouri & O'Toole, 2005; Lambert & Sockett, 2008) in DNB growth medium, following the Section 3.2

3.2 Preparation of *B. bacteriovorus* suspension for predatory assays (see Note 7)

1. Two small pieces of agar are removed with a inoculating loop volume 1 μL from a commercially available "double-layer agar plate" of *B. bacteriovorus* preying on *P. aeruginosa*, as recommended by DSMZ instructions, and added to 60 mL of DNB growth medium in an Erlenmeyer flask (250 mL type)
2. Incubate at 30 °C under agitation (180 rpm) for 48 h (see **Note 8**)
3. Upon reaching 48 h, pass 30 mL of the clear lysate (15 mL + 15 mL, thus in two steps) by means of a 20 mL syringe through a 0.45 μm-filter membrane once in order to remove prey cells and to not excessively damage the predator flagellum
4. Put the 30 mL 0.45 μm-filtrate in two separate 15 mL tubes, and centrifuge the two tubes (balance the rotor) at 10,000*g* for 30 min
5. Resuspend the obtained pellet in 10 mL of TSB (see **Note 9**)
6. In order to exclude carryover of the prey (*P. aeruginosa*), plate 100 μL of the filtrate on TSA agar plates by means of L-shaped loops and incubate at 37 °C overnight. This step has to be performed in parallel with the preceding one: if any prey colony is visible on TSA plates, the obtained suspension of *B. bacteriovorus* cannot be used

3.3 Predation assays on prey cultures

1. Pick up one colony of prey (*P. aeruginosa*) (see **Note 4**) from TSA plates and grow it overnight in 20 mL (use a 50 mL plastic tube) of TSB at 37 °C, 200 rpm
2. Centrifuge bacterial culture (5000*g*, 15 min) and suspend pellet in pre-warmed TSB (37 °C) till reaching an $OD_{600} = 1$ (it should be around 30 mL, use a spectrophotometer to ascertain the right volume)
3. Use 100 μL of suspended culture to inoculate 60 mL of pre-warmed TSB in an Erlenmeyer flask (250 mL type), and then incubate at 30 °C with shaking (200 rpm)
4. Follow bacterial growth each hour with a spectrophotometer, and, at $OD_{600} = 1$ split the prey culture into three different 100 mL-flasks at equal volumes (20 mL): (1) the first one left as it is; (2) the second

one added with 2 mL of *B. bacteriovorus* suspension (see **Note 10**); (3) the third one added with 2 mL of *B. bacteriovorus* suspension 0.22 μm-filtered (see **Note 11**).

5. Incubate flasks at 30 °C with shaking (200 rpm) into the same incubator/shaker (see **Note 12**)
6. Assess prey levels (*P. aeruginosa*) plating 100 μL taken from the above-mentioned first and second flasks every 2 h, on two different TSA plates (see **Note 13**)

3.4 Predatory activity of *B. bacteriovorus* on static biofilms

1. Dilute prey (*P. aeruginosa*) overnight cultures in TSB till reaching an $OD_{600} = 1$
2. Use 200 μL of the dilution to inoculate a 48-well plate, followed by incubation at 37 °C for 24 h (see **Note 14**)
3. After removing the exhausted media, remove eventually present planktonic bacteria by washing, using once 100 μL of Phosphate Buffered Saline (PBS) solution (see **Note 15**)
4. After washing, add 200 μL of *B. bacteriovorus* preparation (see "Preparation of *B. bacteriovorus* for predatory assays" paragraph) to ½ plate, while ¼ is added with 200 μL of 0.22 μm-filtered *B. bacteriovorus* preparation, and another ¼ is added with 200 μL of TSB (taken as control)
5. Incubate plates at 37 °C for additional 24 h, in order to allow *B. bacteriovorus* predation against prey biofilm
6. Wash gently all wells with PBS (100 μL) for three times (see **Note 15**)
7. Add 100 μL of 1% crystal violet solution to each well and wait for 5 min
8. Eliminate the dye in excess by gentle washing three times with PBS (100 μL) (see **Note 15**)
9. Dry the plate in a thermostat (80 °C) for 5 min
10. Add 250 μL of 33% glacial acetic acid to each well and wait for 15 min
11. Read the plate at OD_{570} with a Microplate Reader (see **Note 16**)

3.5 PCR procedure work-flow

The following methods aim to evaluate, by quantitative PCR technique (qPCR), the presence of *B. bacteriovorus* in DNA samples obtained from different sources (feces, intestinal biopsies, wastewater, etc.). For this purpose, carry out the following PCR reactions on the total DNA in order to verify the presence of *B. bacteriovorus*: (i) *Bdellovibrio* species-specific end-point PCR; (ii) *Bdellovibrio* species-specific qPCR.

3.5.1 Detection of B. bacteriovorus by end-point PCR

1. *B. bacteriovorus* specific primers Bd529F (5′-GGTAAGACGAGGGA TCCT-3′) and Bd1007R (5′-TCTTCCAGTACATGTCAAG-3′) are used to amplify a 481-bp trait of the 16S rDNA gene (see **Note 17**)
2. At the same time, primers targeting a 910-bp trait of the *hit* locus, specific for *B. bacteriovorus*, are employed: Hit_FW (5′-GACAGATGG GATTACTGTCTTCC-3′) and Hit_RW (5′-GTGTGATGACGA CTGTGAACGG-3′) (see **Note 17**)
3. PCR reaction (25 μL) contains 1× buffer for PCR, 300 μg/mL bovine serum albumin (BSA), 2.5 mM $MgCl_2$, 200 μM for each dNTP, 0.5 μM of each primer, 1.25 U of Taq polymerase and 100 ng of normalized total DNA. Use the 96-well plate avoiding the external wells in order to ensure reliability of results (see **Notes 18, 19**)
4. For 16S rDNA trait, sample DNA is amplified under the following conditions: 95 °C for 5 min, 20 cycles of 95 °C for 1′, 53 °C for 1′, 72 °C for 1′, and a final step of 72 °C for 10 min
5. For *hit* locus, sample DNA is amplified under the following conditions: 95 °C for 5 min, 20 cycles of 95 °C for 1′, 60 °C for 1′, 72 °C for 1′, and a final step of 72 °C for 10 min
6. Prepare three individual PCR reactions for each sample, then unify them to achieve a final volume of 75 μL
7. Concentrate the unified PCR volume (75 μL) with SpeedVac to reach a final volume approximately equal to 1/3 of the original (25 μL), hereafter named "concentrated PCR" (see **Note 20**)
8. Run 20 μL of the concentrated PCR on an 1% agarose gel containing GelRed 1×, for 1 h at 80 V (see **Note 21**)
9. Put the gel on a transilluminator, photograph it, and perform densitometry analysis with Phoretix 1D software, based on a marker lane with bands of known DNA length and quantity (see **Note 22**)
10. Alternatively, measure DNA concentration by NanoDrop spectrophotometer at 260 nm, using 1 mL of the concentrated PCR (see **Note 22**)
11. Normalize the quantity obtained with densitometry or NanoDrop depending on the target amplified: divide by two for the 16S rDNA, while leave unaltered for the *hit* locus (see **Note 23**)

3.5.2 Detection of B. bacteriovorus by quantitative PCR (see Note 24)

1. Primers used are Bd347F (3′-GGAGGCAGCAGTAGGGAATA-5′) and Bd549R (5′-GCTAGGATCCCTCGTCTTACC-3′), while the probe is the Bb396P (5′FAM-TTCATCACTCACGCGGCGTC-TAMRA3′)
2. The qPCR reaction mix is made up with 10 μL of "2X qPCR ProbesMaster" from Jena Bioscience GmbH (composition: qPCR Pol, dATP, dCTP, dGTP, dUTP, reaction buffer with KCl, $(NH_4)_2SO_4$ and $MgCl_2$, ROX, stabilizers), then 900 nM for both of the primers, 50 nM for the probe, and 5 μL of template (corresponding to 200 ng of normalized total DNA) (see **Notes 18, 19**)
3. Adjust the final reaction volume to 20 μL with PCR-grade distilled water
4. Thermal cycling conditions are: 2 min at 95 °C (initial denaturation), followed by 50 repeats of 15 s at 95 °C (denaturation) and 1 min at 60 °C (annealing and extension). Collect data during the annealing phase
5. To construct the standard curve, quantify the pUC57 plasmid containing the 16 rDNA trait insert by spectroscopy at 260 nm, and 10-fold serially dilute it in the range of 10^9–10^0 copies (Vandecasteele et al., 2002) (see **Note 25**)
6. Results of qPCR are normalized for the number of 16S rDNA operons (two within *B. bacteriovorus* genome) (Van Essche et al., 2009), and expressed as "number of genome copies/mg of sample" (see **Note 26**)

4. Notes

1. The catalog number and supplier are provided for reference, but a comparable product is available from multiple sources at a comparable price. Regarding softwares, even open-source items could be utilizable.
2. It is strongly advised to use nuclease-free plasticware, reagents, and filtered tips.
3. The strain has to be bought in the actively growing culture on a "double-layer agar plate" containing diluted Nutrient-broth (NB) (1:10 dilution of NB amended with 3 mM $MgCl_2 \cdot 6H_2O$ and 2 mM $CaCl_2 \cdot 2H_2O$ [pH 7.2]) and agar (0.6% upper layer) with enclosed prey cells of *P. aeruginosa*. *B. bacteriovorus* forming lysis plaque is well recognizable (Starr, 1975).

4. The strain could be also *Escherichia coli*, or another Gram-negative species strain, and it has to be bought in freeze dried form (DSMZ or ATCC collection).
5. To obtain an enriched predator preparation to be used in predation assays, modify the cultivation media doubling the concentration of NB from 0.8 to 1.6 g/L.
6. Assess strain identification again with Vitek2 or MALDI-TOF in order to ascertain that no environmental contaminants of other bacterial strains were introduced.
7. *B. bacteriovorus* suspension has to be prepared fresh each time for subsequent experiments.
8. Through microscopic observations at 100× magnification and the hanging drop technique, follow and measure every 2 h the growth of *B. bacteriovorus*, discernible by a reduction in OD_{600} turbidity (clear lysate).
9. This step is necessary to have a three-times concentrated suspension of *B. bacteriovorus* and to remove the DNB broth, which initial tests showed to interfere with prey biofilm (data not shown).
10. See "*B. bacteriovorus* suspension for predatory assays" paragraph.
11. The third flask is used for assessing the action of *B. bacteriovorus*-derived lytic enzymes eventually released into the medium.
12. Bacterial growth is measured every hour in two different ways: (1) turbidometric method with optical density OD_{600} (BioPhotometer, Eppendorf); (2) bright field through the "hanging drop" technique (DM4 B Upright Microscope, Leica Microsystems).
13. Employ appropriate serial dilutions in order to ease colony counting. Colonies formed on plates after 17 h at 37 °C could be automatically counted by means of the TotalLab TL120 software (Non-linear Dynamics), "colony counting" module. Predation assays in TSB broth are performed in triplicate: OD_{600} values and colony forming units (CFU) per mL are expressed as mean ± SD of the mean. Graphs and statistical tests are done with Prism 5 software (GraphPad, La Jolla, California, USA).
14. This step is necessary to form a stable prey biofilm.
15. PBS washing needs to be carried out gently without tipping the bottom of the well or disrupting the biofilm.
16. Wilcoxon Signed Rank test is employed to assess differences in biofilm amount, and a *P*-value less than or equal to 0.05 is considered

statistically significant. Graphs and statistical tests are done with Prism 5 software (GraphPad, La Jolla, California, USA).

17. Two sets of primers are used because they are reported to be specific for *B. bacteriovorus* PCR amplification from total DNA (Cotter & Thomashow, 1992; Davidov et al., 2006; Schwudke et al., 2001).

18. Appropriate positive (*B. bacteriovorus* strain HD100) and negative (water) controls need to be employed.

19. In order to minimize the PCR bias, and to avoid the problem due to the expected low levels of *B. bacteriovorus* within samples, perform three individual PCR reactions for each sample. Thus, three PCR reactions for 16S rDNA trait, then other three PCR reactions for the *hit* locus.

20. Stop the SpeedVac concentration every 30 min to control some chosen samples for the right amount of volume obtained. Use the 100 μL pipette in this way: keep all of the volume of a defined sample, then rotate the ring selector on the top of the pipette until the liquid reaches the top of the tip without leaking from it, then the obtained reading is the actual volume of the sample.

21. Agarose gel could be made with TAE (Tris-acetate-EDTA) or TBE (Tris-borate-EDTA) depending on the buffer used. It is advised to not combine TAE with TBE and vice versa. GelRed could be used within the gel (advised) or for post-run coloration.

22. The results obtained by densitometry analysis and Nanodrop measures are in agreement one each other.

23. Ensure that both primer pairs used, for 16S rDNA trait and *hit* locus, are targeted towards the *B. bacteriovorus* genome. Perform an in silico PCR method through the online tool Primer-BLAST present at the National Center for Biotechnology Information (NCBI) website of the University of the National Library of Medicine in Washington (https://www.ncbi.nlm.nih.gov/tools/primer-blast/index.cgi), employing the sequenced genome of the strain HD100 (RefSeq NC_005363). Determine that *B. bacteriovorus* has two 16S rDNA operons (first target, from 820,075 bp to 820,555 bp, and second target, from 1,688,139 bp to 1,688,619 bp), while it harbors only one *hit* locus on the reverse strand (from 96,861 bp to 97,770 bp). Use the obtained information to properly normalize the relative abundances of *B. bacteriovorus*.

24. A quantitative PCR (qPCR) approach with standard curve is implemented in order to assay the absolute abundances of *B. bacteriovorus* in our samples. Primers and hydrolysis probe used are taken from the literature (Van Essche et al., 2009), amplifying a *B. bacteriovorus*-specific region of 16S rDNA.

25. A pUC57 plasmid (or other with similar characteristics) is needed to replicate, albeit at a certain extent, the tridimensional conformation of the genomic DNA, which are important for the sake of qPCR reproducibility, especially in terms of Taq accessibility. Such an approach is not reproducible with PCR products as a starting sample for serial dilution in order to build the standard curve, because PCR products lack the bidimensional and the tridimensional folding of DNA. The mean reaction efficiency (E) of the standard curve, calculated from the mean value of the slope by means of the eq. $E = 10(-1/\text{slope})$, is $99.4\% \pm 1.6\%$. The lowest reproducible detection level of the qPCR is 10 plasmids per reaction, each containing one target sequence.

26. Please refer to the Minimum Information for publication of Quantitative real-time PCR Experiments (MIQE) guidelines (Bustin, 2024; Bustin et al., 2009).

5. Concluding remarks

BALOs are of fundamental importance within the ecological equilibria of a microbial community to understand the tiny interrelationships among bacterial members (Cohen et al., 2021; Johnke et al., 2020; Martin, 2002) and, potentially, with the eukaryotic counterpart (Gupta et al., 2016). Knowing how to grow, propagate, detect and assess predatory activities of BALOs would be amenable to new achievements in this fantastic research topic, which is older than 80 years (Pérez et al., 2016).

References

Balows, A., et al. (2013). *The prokaryotes: A handbook on the biology of bacteria: Ecophysiology, isolation, identification*. Applications Springer Science & Business Media.

Bremer, N., Tria, F. D. K., Skejo, J., Garg, S. G., & Martin, W. F. (2022). Ancestral state reconstructions trace mitochondria but not phagocytosis to the last eukaryotic common ancestor. *Genome Biology and Evolution*, *14*(6), evac079. https://doi.org/10.1093/gbe/evac079.

Burnham, J. C., et al. (1968). Electron microscopic observations on the penetration of Bdellovibrio bacteriovorus into gram-negative bacterial hosts. *Journal of Bacteriology*, *96*, 1366–1381.

Bustin, S. A. (2024). Improving the quality of quantitative polymerase chain reaction experiments: 15 years of MIQE. *Molecular Aspects of Medicine*, *96*, 101249.

Bustin, S. A., et al. (2009). The MIQE guidelines: Minimum information for publication of quantitative real-time PCR experiments. *Clinical Chemistry*, *55*, 611–622.
Casida, L. E., Jr. (1988). Minireview: Nonobligate bacterial predation of bacteria in soil. *Microbial Ecology*, *15*, 1–8.
Caulton, S. G., et al. (2024). Bdellovibrio bacteriovorus uses chimeric fibre proteins to recognize and invade a broad range of bacterial hosts. *Nature Microbiology*, *9*, 214–227.
Cohen, Y., et al. (2021). Community and single cell analyses reveal complex predatory interactions between bacteria in high diversity systems. *Nature Communications*, *12*, 5481.
Comandatore, F., et al. (2021). Modeling the life cycle of the intramitochondrial bacterium "Midichloria mitochondrii" using electron microscopy data. *MBio*, *12*, e0057421.
Cotter, T. W., & Thomashow, M. F. (1992). Identification of a Bdellovibrio bacteriovorus genetic locus, hit, associated with the host-independent phenotype. *Journal of Bacteriology*, *174*, 6018–6024.
Davidov, Y., & Jurkevitch, E. (2009). Predation between prokaryotes and the origin of eukaryotes. *BioEssays*, *31*, 748–757.
Davidov, Y., et al. (2006). Structure analysis of a soil community of predatory bacteria using culture-dependent and culture-independent methods reveals a hitherto undetected diversity of Bdellovibrio-and-like organisms. *Environmental Microbiology*, *8*, 1667–1673.
Esteve, I., & Gaju, N. (1999). Bacterial symbioses. Predation and mutually beneficial associations. *International Microbiology*, *2*, 81–86.
Esteve, I., et al. (1983). Electron microscope study of the interaction of epibiontic bacteria with Chromatium minus in natural habitats. *Microbial Ecology*, *9*, 57–64.
Fallon, R. D., & Brock, T. D. (1979). Lytic organisms and photooxidative effects: Influence on blue-green algae (cyanobacteria) in Lake Mendota, Wisconsin. *Applied and Environmental Microbiology*, *38*, 499–505.
Guerrero, R., et al. (1986). Predatory prokaryotes: Predation and primary consumption evolved in bacteria. *Proceedings of the National Academy of Sciences of the United States of America*, *83*, 2138–2142.
Gupta, S., et al. (2016). Effect of predatory bacteria on human cell lines. *PLoS One*, *11*, e0161242.
Horowitz, A. T., et al. (1974). Growth cycle of predacious Bdellovibrios in a host-free extract system and some properties of the host extract. *Journal of Bacteriology*, *117*, 270–282.
Iebba, V., et al. (2013). Higher prevalence and abundance of Bdellovibrio bacteriovorus in the human gut of healthy subjects. *PLoS One*, *8*, e61608.
Iebba, V., et al. (2014). Bdellovibrio bacteriovorus directly attacks Pseudomonas aeruginosa and Staphylococcus aureus cystic fibrosis isolates. *Frontiers in Microbiology*, *5*, 280.
Johnke, J., et al. (2020). Bdellovibrio and like organisms are predictors of microbiome diversity in distinct host groups. *Microbial Ecology*, *79*, 252–257.
Jurkevitch, E. (2006). *Predatory prokaryotes: Biology*. Ecology and Evolution Springer Science & Business Media.
Jurkevitch, E., & Mitchell, R. J. (2020). *The ecology of predation at the microscale*. Springer Nature.
Jurkevitch, E., & Mitchell (Professor), R. J. (2020). *Ecology of predation at the microscale*. Springer. 210 pp.
Kadouri, D., & O'Toole, G. A. (2005). Susceptibility of biofilms to Bdellovibrio bacteriovorus attack. *Applied and Environmental Microbiology*, *71*, 4044–4051.
Kamada, S., Wakabayashi, R., & Naganuma, T. (2023). Phylogenetic revisit to a review on predatory bacteria. *Microorganisms*, *11*(7), 1673. https://doi.org/10.3390/microorganisms11071673.
Kessel, M., & Shilo, M. (1976). Relationship of Bdellovibrio elongation and fission to host cell size. *Journal of Bacteriology*, *128*, 477–480.

Lambert, C., & Sockett, R. E. (2008). Laboratory maintenance of Bdellovibrio. *Current Protocol in Microbiology, Chapter* 7, Unit 7B.2.

Lambert, C., et al. (2003). A novel assay to monitor predator-prey interactions for Bdellovibrio bacteriovorus 109 J reveals a role for methyl-accepting chemotaxis proteins in predation. *Environmental Microbiology, 5*, 127–132.

Lambert, C., et al. (2009). A predatory patchwork: Membrane and surface structures of Bdellovibrio bacteriovorus. *Advances in Microbial Physiology, 54*, 313–361.

Lambina, V. A., et al. (1987). Importance of Bdellovibrio in regulating microbial cenoses and self-purification processes in domestic sewage. *Mikrobiologiia, 56*, 860–864.

Lin, D., & McBride, M. J. (1996). Development of techniques for the genetic manipulation of the gliding bacteria Lysobacter enzymogenes and Lysobacter brunescens. *Canadian Journal of Microbiology, 42*, 896–902.

Marchi, S., et al. (2022). Control of host mitochondria by bacterial pathogens. *Trends in Microbiology, 30*, 452–465.

Markelova, N. Y. (2010). Predacious bacteria, Bdellovibrio with potential for biocontrol. *International Journal of Hygiene and Environmental Health, 213*, 428–431.

Martin, M. O. (2002). Predatory prokaryotes: An emerging research opportunity. *Journal of Molecular Microbiology and Biotechnology, 4*, 467–477.

Maurice, N. M., & Sadikot, R. T. (2023). Mitochondrial dysfunction in bacterial infections. *Pathogens, 12*(8), 1005. https://doi.org/10.3390/pathogens12081005.

Núñez, M. E., et al. (2003). Investigations into the life cycle of the bacterial predator Bdellovibrio bacteriovorus 109J at an interface by atomic force microscopy. *Biophysical Journal, 84*, 3379–3388.

Pantanella, F., et al. (2018). Behaviour of Bdellovibrio bacteriovorus in the presence of gram-positive Staphylococcus aureus. *The New Microbiologica, 41*, 145–152.

Pérez, J., et al. (2016). Bacterial predation: 75 years and counting! *Environmental Microbiology, 18*, 766–779.

Sacchi, L., et al. (2004). A symbiont of the tick Ixodes ricinus invades and consumes mitochondria in a mode similar to that of the parasitic bacterium Bdellovibrio bacteriovorus. *Tissue & Cell, 36*, 43–53.

Schwudke, D., et al. (2001). Taxonomic studies of predatory bdellovibrios based on 16S rRNA analysis, ribotyping and the hit locus and characterization of isolates from the gut of animals. *Systematic and Applied Microbiology, 24*, 385–394.

Shilo, M., & Bruff, B. (1965). Lysis of Gram-negative bacteria by host-independent ectoparasitic Bdellovibrio bacteriovorus isolates. *Journal of General Microbiology, 40*, 317–328.

Singh, R. P., et al. (2022). *Microbes in microbial communities: Ecological and applied perspectives*. Springer Nature.

Sockett, R. E. (2009). Predatory lifestyle of Bdellovibrio bacteriovorus. *Annual Review of Microbiology, 63*, 523–539.

Sockett, R. E. (2023). Learning with Bdellovibrio. *Nature Microbiology, 8*, 1189–1190.

Spier, A., Stavru, F., & Cossart, P. (2019). Interaction between intracellular bacterial pathogens and host cell mitochondria. *Microbiology Spectrum*, 7. https://doi.org/10.1128/microbiolspec.bai-0016-2019.

Starr, M. P. (1975). *Bdellovibrio* as symbiont; the associations of Bdellovibrios with other bacteria interpreted in terms of a generalized scheme for classifying organismic associations. *Symposia of the Society for Experimental Biology*, 93–124.

Stavru, F., et al. (2020). When bacteria meet mitochondria: The strange case of the tick symbiont Midichloria mitochondrii. *Cellular Microbiology, 22*, e13189.

Stolp, H., & Starr, M. P. (1963). *Bdellovibrio bacteriovorus* gen. et sp. n., a predatory, ectoparasitic and bacteriolytic microorganism. *Antonie Van Leeuwenhoek, 29*, 217–248.

Uzum, Z., et al. (2023). Three-dimensional images reveal the impact of the endosymbiont Midichloria mitochondrii on the host mitochondria. *Nature Communications*, *14*, 4133.

Van Essche, M., et al. (2009). Development and performance of a quantitative PCR for the enumeration of Bdellovibrionaceae. *Environmental Microbiology Reports*, *1*, 228–233.

Vandecasteele, S. J., et al. (2002). Use of gDNA as internal standard for gene expression in staphylococci in vitro and in vivo. *Biochemical and Biophysical Research Communications*, *291*, 528–534.

Assessment of adhering and invading properties of *Escherichia coli* strains

Valerio Iebba*
Gustave Roussy Cancer Campus, Villejuif, France
*Corresponding author: e-mail address: viebba@units.it

Contents

Abstract

Gastrointestinal infections, caused by Enterobacteriaceae, pose a major global health challenge, resulting in significant morbidity and mortality. Enhanced adherence and invasion properties are widespread among enteric pathogenic species, particularly those linked to invasive infections such as some pathovars of *Escherichia coli* or pathogens like *Shigella* and *Salmonella*. Pathogenic *E. coli* strains are categorized into various pathotypes, including diarrheagenic *E. coli* (DEC) and extraintestinal pathogenic *E. coli* (ExPEC). Notably, Enteroinvasive *E. coli* (EIEC) and Adherent-invasive *E. coli* (AIEC) demonstrate significant invasive properties. EIEC, similar to *Shigella*, invades intestinal epithelial cells causing dysentery-like illness, while AIEC persists in the gut epithelium, potentially contributing to chronic inflammatory bowel diseases (IBD). Techniques like cell culture assays are vital for assessing *E. coli*'s adherence and invasion capabilities, with specific virulence factors such as fimbriae and type III secretion systems (T3SS) playing crucial roles. Comparatively, *Shigella* and *Salmonella* also utilize T3SS for epithelial cell invasion, but with distinct effector proteins and mechanisms. Understanding these differences is crucial for diagnosis and treatment, as advanced molecular diagnostics

Methods in Cell Biology, Volume 194
ISSN 0091-679X
https://doi.org/10.1016/bs.mcb.2024.08.011

improve the identification of invasive *E. coli* strains. Potential therapeutic interventions targeting fimbrial adherence, T3SS and effector proteins offer promising avenues for developing antivirulence drugs. Here are provided protocols for studying the adherence and invasion properties of *E. coli* and other Enterobacteriaceae to enhance diagnostic methods, ultimately improving the management of enteric infections.

1. Introduction

Gastrointestinal infections are a significant global health concern, causing substantial morbidity and mortality (Kaper, Nataro, & Mobley, 2004). Enterobacteriaceae, a large family of Gram-negative bacteria, are frequently implicated in these infections. While some strains of Enterobacteriaceae are commensals of the human gut microbiota, others have evolved pathogenic characteristics, causing a spectrum of diseases ranging from mild diarrhea to life-threatening dysentery (Kotloff et al., 2013). Among these pathogens, *Escherichia coli* strains hold particular attention (Gomes, Dobrindt, Farfan, & Piazza, 2021). Traditionally, *E. coli* pathotypes were categorized based on their toxin production and clinical presentation (Law, Gur-Arie, Rosenshine, & Finlay, 2013; Rivas, Mellor, Gobius, & Fegan, 2015). However, recent research has highlighted the significance of epithelial cell adhesion and invasion as key virulence factors for certain *E. coli* strains (Buc et al., 2013; Johnson, 2017; Kaper et al., 2004), such as for other prominent enteric pathogens like *Shigella* and *Salmonella*. *E. coli* encompasses a diverse group of strains, with some residing harmlessly within the human gut microbiota and others causing a variety of intestinal and extraintestinal diseases (Croxen et al., 2013). Pathogenic *E. coli* strains are classified into pathotypes based on their virulence factors and clinical syndromes they cause (Garcia-Angulo, Farfan, & Torres, 2013; Robins-Browne et al., 2016). These include diarrheagenic *E. coli* (DEC), extraintestinal pathogenic *E. coli* (ExPEC). Two pathotypes are particularly relevant when discussing invasive properties: Enteroinvasive *E. coli* (EIEC) and Adherent-invasive *E. coli* (AIEC) (Boudeau, Glasser, Masseret, Joly, & Darfeuille-Michaud, 1999; Costa et al., 2020; Di Pasquale et al., 2016; Martinez-Medina & Garcia-Gil, 2014; Schippa et al., 2012). EIEC strains share similarities with Shigella species, exhibiting a high degree of intestinal epithelial cell invasion and causing dysentery-like illness (The, Thanh, Holt, Thomson, & Baker, 2016). AIEC strains, on the other hand, demonstrate a unique ability to adhere and persist within the gut epithelium, potentially contributing to chronic inflammatory bowel diseases (IBD) (Boudeau et al., 1999;

Darfeuille-Michaud et al., 2004; Iebba et al., 2012; Keita et al., 2020; Nguyen et al., 2014; Schippa et al., 2009; Schippa et al., 2012). Assessing the adhering and invading properties of *E. coli* strains is crucial for understanding their pathogenic potential. Several techniques can be employed, including cell culture assays, molecular analysis of virulence genes, and organoids (Camprubí-Font, Ewers, Lopez-Siles, & Martinez-Medina, 2019; Camprubí-Font & Martinez-Medina, 2020; Mayorgas et al., 2021). Cell culture assays, such as the gentamicin protection assay, allow researchers to evaluate the ability of *E. coli* strains to adhere to and invade epithelial cell lines (Camprubí-Font et al., 2019; Mayorgas et al., 2021; Sharma & Puhar, 2019). In-vivo models using mice or zebrafish can further elucidate the disease process and host response (Law et al., 2013). Additionally, analysis of genes associated with adherence and invasion, such as those encoding fimbriae (adhesion pili) and type III secretion systems (T3SS), provides valuable insights into the molecular mechanisms employed by *E. coli* strains (Iebba et al., 2012; Camprubí-Font et al., 2019).

E. coli strains utilize a diverse arsenal of virulence factors to facilitate their adherence and invasion of epithelial cells (Robins-Browne et al., 2016). Fimbriae, hair-like structures on the bacterial surface, play a critical role in the initial attachment to host cells (Iebba et al., 2012; Camprubí-Font et al., 2019). Specific fimbrial types, such as type 1 fimbriae (FimH) and P fimbriae, have been associated with adherence of EIEC and AIEC strains (Garcia-Angulo et al., 2013; Robins-Browne et al., 2016). Once adherence is established, *E. coli* strains often employ T3SS to inject effector proteins into the host cell cytoplasm (Clements, Berger, Lomma, & Frankel, 2013; Worrall, Bergeron, & Strynadka, 2013). These effector proteins manipulate host cell processes, allowing for bacterial uptake, actin cytoskeleton rearrangements, and disruption of barrier functions (Clements et al., 2013; Worrall et al., 2013). EIEC strains utilize T3SS-mediated effector proteins, such as IcsA and Tir, to induce actin polymerization and promote bacterial entry into the host cell. AIEC strains, on the other hand, exhibit a more subtle invasion process, utilizing outer-membrane vesicles (OMVs) and long polar fimbriae to establish persistent intracellular niches within the gut epithelium (Lee et al., 2019; Rolhion et al., 2010; Rolhion, Hofman, & Darfeuille-Michaud, 2011).

While *E. coli* strains, particularly EIEC and AIEC, are increasingly recognized for their invasion capabilities, other Enterobacteriaceae also possess significant virulence factors associated with epithelial cell adhesion and invasion. *Shigella* species, the classic causative agents of bacillary dysentery, exhibit a highly invasive nature (Lampel, Formal, & Maurelli, 2018; Sharma & Puhar, 2019). *Shigella* utilizes a T3SS to inject effector proteins like IpaA and IcsA, which manipulate host cell processes and trigger actin polymerization, enabling bacterial entry (Mattock & Blocker, 2017). *Salmonella enterica* serovars, another prominent group of enteric pathogens, also possess virulence factors facilitating epithelial cell invasion (Eng et al., 2015; Jajere, 2019). *Salmonella* employs a complex T3SS-dependent invasion process involving effector proteins like SipA and SopB (Srikanth, Mercado-Lubo, Hallstrom, & McCormick, 2011). These effectors promote bacterial uptake and intracellular survival within specialized vacuoles, allowing *Salmonella* to persist within the host and cause systemic infections (Srikanth et al., 2011; Zhou et al., 2023).

By comparing the virulence factors employed by *E. coli*, *Shigella*, and *Salmonella*, we can gain a deeper understanding of their diverse strategies for epithelial cell invasion (Du et al., 2016). While all three pathogens utilize T3SS as a core mechanism, the specific effector proteins and their functions differ (Pinaud, Sansonetti, & Phalipon, 2018). EIEC and *Shigella* share some similarities, such as the use of IcsA for actin polymerization and bacterial entry. However, EIEC strains often exhibit a less aggressive invasion compared to *Shigella*, which may contribute to the varying clinical presentations of their infections (Clements et al., 2013; Worrall et al., 2013). *Salmonella*, on the other hand, employs a more complex invasion process involving a wider array of effector proteins like SipA and SopB. This complex interplay allows *Salmonella* to not only invade epithelial cells but also establish intracellular vacuoles for persistence within the host (Srikanth et al., 2011; Zhou et al., 2023). Additionally, *E. coli* strains, particularly AIEC, demonstrate a unique ability to establish chronic infections through persistent adherence within the gut epithelium (Darfeuille-Michaud et al., 2004; Iebba et al., 2012; Keita et al., 2020). This strategy differs from the more acute invasive processes observed in EIEC, *Shigella*, and *Salmonella*, even involving the production of outer-membrane vesicles (OMVs) (Rolhion et al., 2010; Rolhion et al., 2011). Furthermore, the differences in how these pathogens manipulate host cell processes highlight potential targets for therapeutic interventions. Disrupting T3SS function or effector protein activity could be promising strategies for developing novel antivirulence drugs against *E. coli*, *Shigella*, and *Salmonella* (Clements et al., 2013; Worrall et al., 2013). Additionally, understanding the unique persistence mechanisms of AIEC strains could lead to novel therapeutic approaches for chronic inflammatory bowel diseases (Camprubí-Font et al., 2019; Mayorgas et al., 2021).

Mitochondria, central to cellular metabolism, apoptosis, and immune responses, are key targets for pathogens like *Escherichia coli*, *Shigella*, and *Salmonella*. These bacteria, among others, have evolved sophisticated mechanisms to manipulate mitochondrial functions, enhancing their survival and proliferation within host cells (Marchi, Morroni, Pinton, & Galluzzi, 2022; Tiku, Tan, & Dikic, 2020). Enteropathogenic *Escherichia coli* (EPEC) utilizes type-three secretion system (T3SS) effectors, such as Map and EspF, which are directed to mitochondria through mitochondrial targeting sequences (MTSs) (Nougayrède & Donnenberg, 2004; Ramachandran et al., 2020). Once inside the mitochondria, these effectors cause disruptions in mitochondrial morphology, interfere with calcium homeostasis, and trigger apoptosis. In contrast, the T3SS effectors IpaD from *Shigella flexneri* (Arizmendi, Picking, & Picking, 2016) and SipD from *Salmonella typhimurium* do not possess MTSs but still initiate cell-death pathways (Arizmendi et al., 2016; Hernandez, Pypaert, Flavell, & Galán, 2003). Specifically, IpaD causes mitochondrial depolarization, leading to the activation of the intrinsic apoptosis pathway, while SipD induces mitochondrial damage and triggers non-canonical autophagy-mediated type II programmed cell death. Additionally, many bacteria secrete pore-forming toxins (PFTs), for example colicins from *E. coli* and cytolysin A from *E. coli* and *Salmonella enterica*, able to induce calcium influx into cells leading to a subsequent mitochondrial dysfunction and apoptosis (Verma, Gandhi, Lata, & Chattopadhyay, 2021).

Understanding the nuances in adhering and invading properties of different Enterobacteriaceae has crucial implications for diagnosis and treatment of enteric infections. Distinguishing between *E. coli* pathotypes, particularly EIEC and AIEC, could be achieved using traditional diagnostic methods (Camprubí-Font et al., 2019; Mayorgas et al., 2021; O'Brien et al., 2017) For example, amplifying the *ipaH* gene, which is found in multiple copies on both the *inv* plasmid and the chromosome, serves as a diagnostic marker for the detection of EIEC and Shigella (Pakbin, Brück, & Rossen, 2021; van den Beld & Reubsaet, 2012). Recent advancements in molecular diagnostics, such as real-time PCR assays (Barbau-Piednoir et al., 2018; Vashisht et al., 2023), or the whole-genome sequencing (WGS) of isolated strains, offer improved accuracy in identifying these invasive *E. coli* strains

(Quainoo et al., 2017; Zhang et al., 2015). Unlike other *E. coli* pathotypes, AIEC strains lack typical virulence genes but share genetic similarities with extraintestinal pathogenic *E. coli* (ExPEC) (Boudeau et al., 1999; Glasser et al., 2001; Martinez-Medina et al., 2009; Nash et al., 2010; O'Brien et al., 2017). Identifying molecular markers for AIEC has proven challenging, with no definitive genetic signature identified so far (Bonet-Rossinyol, Camprubí-Font, López-Siles, & Martinez-Medina, 2023; Céspedes et al., 2017; Zangara, Darwish, & Coombes, 2023). Previous studies proposed potential biomarkers, but their accuracy and applicability across diverse *E. coli* strains remain limited (Deshpande, Wilkins, Mitchell, & Kaakoush, 2015; Desilets et al., 2016). Notably, an algorithm based on single nucleotide polymorphisms (SNPs) was developed, achieving an 81% accuracy in detecting AIEC, though it showed regional bias and was less effective for strains from different geographical locations (Camprubí-Font et al., 2018; Camprubí-Font et al., 2020). Recent transcriptomic studies have attempted to uncover the molecular underpinnings of AIEC pathogenicity. For instance, over-expressed genes in AIEC during intestinal epithelial cells infection include those in the *fim* operon, which encodes type 1 fimbriae critical for bacterial adhesion to epithelial cells, a process enhanced in CD patients due to upregulated expression of CEACAM6 (Bonet-Rossinyol et al., 2023; Céspedes et al., 2017; Zangara et al., 2023). However, further research is required to validate these findings across diverse AIEC strains and to establish reliable biomarkers for clinical diagnostics and disease management (Elhenawy, Oberc, & Coombes, 2018).

Here we describe the protocol to assess adhesion and invasion of *E. coli* strains on a cellular model of intestinal epithelial cells (Caco-2), assuming that they are already isolated from natural sources, such as biopsies or fecal samples, and characterized with state-of-the-art techniques (genotyping, phylotyping, MALDI-TOF, MLST—Multi-locus Sequence Typing, WGS—Whole Genome Sequencing, etc.). For completeness, an isolation protocol of *E. coli* strains from biopsies is provided as reference.

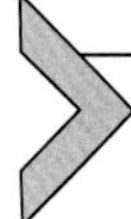

2. Materials

2.1 Common disposables

- 10, 20, 200 and 1000 μL Sartorius filter pipette tips (#Z757764, #Z757772, #Z757799, #Z757810, Merck) (see **Notes 1, 2**)
- 2 mL microcentrifuge tube (#Z628034, Merck) (see **Note 1, 2**)

- 15 mL conical tube (#CLS430766, Merck) (see **Note 1, 2**)
- 50 mL conical tube (#CLS431472, Merck) (see **Note 1, 2**)
- Inoculating loop volume 1 μL (#HS81121A, Merck) (see **Note 1**)
- 24-well plate (#CLS3473, Merck) (see **Notes 1, 2**)
- L-shape spreader (#HS8171B, Merck) (see **Note 1, 2**)
- Petri dishes (size 100 mm × 20 mm) (#P5606, Merck) (see **Note 1, 2**)

2.2 Cells and reagents

- Tryptone Soy Agar (TSA) (#1.46004, Merck) (see **Note 1**)
- MacConkey Agar (#M7408, Merck) (see **Note 1**)
- Tryptone Soy Broth (TSB) (#STBMTSB12, Merck) (see **Note 1**)
- Brain Heart Infusion (BHI) Broth (#53286, Merck) (see **Note 1**)
- Luria-Bertani (LB) broth (#L3022, Merck) (see **Note 1**)
- Phosphate-buffered saline (PBS) (#10010023, Life Technologies) (see **Note 1**)
- Gentamicin Solution 50 mg/mL (#G1397, Merck) (see **Note 1**)
- DL-Dithiothreitol solution (DTT) (#646563, Merck) (see **Note 1**)
- Sterile-filtered water, suitable for cell culture (#W3500, Merck) (see **Note 1**)
- Microbact Biochemical Identification Kits (#MB1074A, Thermo Fisher Scientific) (see **Note 1**)
- Glycerol (#G9012, Merck) (see **Note 1**)
- Nunc® CryoTubes® (#V7884, Merck) (see **Note 1**)
- Caco-2 cell line (#HTB-37 ATCC) (see **Note 1**)
- DMEM—low glucose (#D6046, Merck) (see **Note 1**)
- L-glutamine (#G7513, Merck) (see **Note 1**)
- MEM Non-essential Amino Acid Solution (NEAA) (#M7145, Merck) (see **Note 1**)
- Heat-inactivated Fetal Bovine Serum (FBS) (#F9665, Merck) (see **Note 1**)
- Penicillin-streptomycin solution (#P4333, Merck) (see **Note 1**)
- Triton™ X-100 (~10% in H_2O) (#93443, Merck) (see **Note 1**)

2.3 Equipment (see Note 1)

- Incubator/shaker (#5000IR, VWR)
- Eppendorf benchtop centrifuge 5810G (#EP5810000420, Merck)
- Vortex-Genie® 2 mixer (#Z258423, Merck)

3. Methods

3.1 *E. coli* strains isolation from intestinal biopsies (see Note 3)

1. Collect a fresh biopsy (ca 15 mg) in a 2 mL tube already containing 500 μL of phosphate-buffered saline (PBS).
2. Wash the biopsy three times with 500 μL phosphate-buffered saline (PBS) containing 0.016% dithiothreitol (DTT) final working concentration in order to remove the mucus layer (see **Note 5**). Each wash needs to be done shaking the tube for 30 s, then centrifuge for 1 min at 2000 rpm and discard the supernatant.
3. After the third wash, discard it and add 500 μL of distilled water to the biopsy in order to promote hypotonic lysis, and vortex for 30 s each 5 min for a total time of 30 min, thus you have to vortex for 6 times within the 30 min time window and centrifuge briefly (little spin, 1 min at 1000 rpm) to recollect the lysate at the bottom (see **Note 6**).
4. All the volume of the cell debris left after hypotonic lysis is plated in 10-fold dilution steps (final volume of 500 μL) onto diverse MacConkey or TSA plates. Prepare at least 3-orders of magnitude dilutions (1:10 with PBS in 2 mL tubes), thus to have four tubes (UM unmodified, 1:10, 1:100, 1:1000, or more dilution factors). Plate 500 μL for each plate with an L-shaped spreader.
5. After 24 h of incubation at 37 °C, randomly select suspicious *E. coli* colonies using an inoculating loop, till 5 per biopsy or more (see **Note 7**), putting a single colony in 5 mL of BHI (or TSB or LB) liquid medium within a 15 mL tube. Incubate overnight at 37 °C with 200 rpm agitation and loosely capped.
6. Subculture the selected colonies onto nutrient agar plate (MacConkey or TSA) and successively identify them by Microbact Biochemical Identification Kits (Thermo Fisher Scientific). API ID systems (Biomerieux), MALDI-TOF and WGS are also valid systems to properly identify *E. coli* species.
7. Once identified as *E. coli*, pick up the colony with a loop and resuspend it in 5 mL of liquid broth (TSB or BHI or LB) contained in a 15 mL tube.
8. Let it grow overnight (ON) at 37 °C in agitation at 200 rpm with the cap loosely attached in order to ensure air flow.
9. In order to make a collection of all isolated strains, add 150 μL of pure glycerol (see **Note 8**) to 850 μL of the ON bacterial suspension already

within a cryotube, in order to achieve a 15% vol/vol of "glycerol stock". Mix well with gentle vortexing (3 s, 1000 rpm) in order to have an homogeneous solution. Prepare 5 tubes of glycerol stocks, for each isolate, and store them at -80 °C (see **Note 9**).

3.2 Adhesion assay

1. For static adhesion assays, seed Caco-2 cells (human colonic adenocarcinoma cells) at a density of $2*10^4$ cells/well in 24-well plates and culture to a complete differentiation for 15 days (postconfluence state) before infection in a 5% CO_2 incubator at 37 °C (see **Note 10**).
2. Bacterial colonies, for maximal fimbrial expression and adhesion, are grown overnight in 5 mL of nutrient broth (BHI or LB or TSB) (see **Note 9**), centrifuged at 5000 rpm for 5 min, and the pellet resuspended in 5 mL of PBS in order to keep the same bacterial concentration obtained during the ON (roughly 10^9 CFU/mL order of magnitude), and subsequently left for 48 h at room temperature.
3. Prepare the bacterial solution for Caco2-infection in the following way: keep 1 mL of the previous bacterial suspension in a 2 mL tube, centrifuge 2 min at 4000 rpm, discard supernatant and resuspend the pellet in 1 mL of pre-warmed Caco-2 growth medium (see **Note 10** for its composition, but without penicillin and streptomycin). This step ensures that bacteria are already within the cell line culture medium, without contamination of residual bacterial growth medium.
4. Infect each monolayer with around $1*10^8$ bacteria/well (thus, put 1 mL in the well prepared mixing 100 μL of the previous bacterial solution with 900 μL of pre-warmed Caco-2 cell growth medium without penicillin and streptomycin) and incubate for 3 h at 37 °C. The multiplicity of infection (MOI) is thus roughly 10:1 (see **Note 11**). Perform a triplicate for each experimental point (time, dose, etc.).
5. After infection, wash the monolayer 3 times with PBS, taking care of not dislodging the monolayer. Use a 1000 pipette with filter tip, and gently push up and down 500 μL of PBS for 4–5 times: this washing has to be repeated 3 times (see **Note 12**).
6. Add 500 μL of 1 % Triton X-100 (diluted in deionized water) to the monolayer in order to lyse cells, let stand 5 min (or longer if needed) (see **Note 13**), homogenize the lysate with the aid of a 1000-μL pipette (gently up and down) and prepare 1:10 dilutions as stated in step 4 of Methods 3.1, in a final volume of 500 μL.

7. Plate the whole volume of dilutions (and the unmodified lysate) on different MacConkey or TSA agar plates by means of an L-shaped spreader. Incubate ON at 37 °C.
8. Count the number of colonies in each replicate (for a defined experimental point) to determine the number of CFUs, and take into account the dilution factor to obtain an average count for all the plates for the single infection (the single well in the 24-well plate). In this manner you should have 12 counts (3 replicates, 4 dilutions for each replicate) for each experimental point (time, dose, etc.) to calculate the average count.
9. Define the bacterial adhesion as the "percentage of attached bacteria" compared with the initial inoculum, which is taken as 100%. For example, if the initial inoculum is $I_{initial} = 1*10^8$ bacterial cells and the averaged final count is $I_{final} = 4*10^6$ attached bacterial cells, then the adhesion percentage is $Adh = (I_{final} / I_{initial})*100 = 4\%$, meaning that 4% of the initial inoculum adhered to Caco-2 cells (see **Note 14**). A strain is considered adhesive if its adhesion index is equal or superior to 0.8% of the original inoculum (Darfeuille-Michaud et al., 2004; Iebba et al., 2012; Schippa et al., 2009; Schippa et al., 2012).
10. Another way to determine bacterial adhesion is to define the "mean number of bacteria per cell" (Darfeuille-Michaud et al., 2004; Schippa et al., 2012). Actually, if we have a final number of attached bacterial cells $I_{final} = 4*10^6$ and a final number of Caco-2 cells is $N_{Caco\text{-}2} = 1.28*10^6$ (see **Note 10**), then the mean number of adhering bacteria per cell would be $I_{final} / N_{Caco\text{-}2} = 3.125 \approx 3$. A strain is considered adhesive if its adhesion index is equal or superior to 1 bacteria per cell.

3.3 Invasion assay

1. For the invasion assay, prepare Caco-2 cells and bacterial suspension following the same steps as in adhesion assay (from 3.2.1 till 3.2.5).
2. Add 1 mL of pre-warmed Caco-2 cell culture medium without penicillin and streptomycin, but containing 100 μg/mL of gentamicin (Sigma, St. Louis, MO) (see **Note 15**), and let stand for 1 h to kill extracellular bacteria (see **Note 16**). This procedure is called "gentamicin protection assay" (Falkow, Small, Isberg, Hayes, & Corwin, 1987).
3. After infection, wash the monolayer 3 times with PBS, taking care of not dislodging the monolayer. Use a 1000 pipette with filter tip, and gently push up and down 500 μL of PBS for 4–5 times: this washing has to be repeated 3 times (see **Note 12**).

4. Add 500 μL of 1% Triton X-100 (diluted in deionized water) to the monolayer in order to lyse cells, let stand 5 min (or longer if needed) (see **Note 12**), homogenize the lysate with the aid of a 1000-μL pipette (gently up and down) and prepare dilutions as stated in step 4 of Methods 3.1, in a final volume of 500 μL.
5. Plate the whole volume of dilutions (and the unmodified lysate) on different MacConkey or TSA agar plates by means of an L-shaped spreader. Incubate ON at 37 °C.
6. Count the number of colonies in each replicate (for a defined experimental point) to determine the number of CFUs, and take into account the dilution factor to obtain an average count for all the plates for the single infection (the single well in the 24-well plate). In this manner you should have 12 counts (3 replicates, 4 dilutions for each replicate) for each experimental point (time, dose, etc.) to calculate the average count.
7. Define the bacterial invasion as the "percentage of intracellular bacteria" compared with the initial inoculum, which is taken as 100%. For example, if the initial inoculum is $I_{initial} = 1*10^8$ bacterial cells and the averaged final count of intracellular bacteria is $I_{final} = 2.5*10^6$ internalized bacterial cells, then the invasion percentage is $Inv = (I_{final} / I_{initial})*100 = 2.5\%$, meaning that 2.5% of the initial inoculum entered into Caco-2 cells (see **Note 17**). A strain is considered invasive if its invasion index is equal or superior to 0.1% of the original inoculum (Darfeuille-Michaud et al., 2004; Kittana et al., 2023; Schippa et al., 2012).
8. Another way to determine bacterial invasion is to define the "mean number of intracellular bacteria per cell". Actually, if we have a final number of internalized bacterial cells $I_{final} = 2.5*10^6$ and a final number of Caco-2 cells is $N_{Caco\text{-}2} = 1.28*10^6$ (see **Note 10**), then the mean number of intracellular bacteria per cell would be $I_{final} / N_{Caco\text{-}2} = 1.953 \approx 2$ (see **Note 18**).

4. Notes

1. The catalog number and supplier are provided for reference, but a comparable product is available from multiple sources at a comparable price. Regarding softwares, even open-source items could be utilizable.
2. It is strongly advised to use nuclease-free plasticware, reagents, and filtered tips.

3. The *E. coli* strain could be isolated from several biological (or environmental) samples. Here is provided a simple isolation protocol from intestinal biopsies (colon). Other strains from pathogenic and invasive species (such as *Shigella* or *Salmonella*) could be isolated in the same manner.
4. Assess strain identification again with Vitek2 or MALDI-TOF in order to ascertain that no environmental contaminants of other bacterial strains were introduced.
5. Dithiothreitol (DTT) is used to remove luminal bacteria and the mucus layer, ruling out loosely attached bacteria cells and leaving only the ones attached to the brush border and the intracellular ones. DTT is a reducing agent that loosen the disulphide bonds among mucin proteins, which constitutes the intestinal mucus layer (40–240 μm thick) (Swidsinski et al., 2007)
6. The DTT-washed biopsies could also be crushed (Darfeuille-Michaud et al., 2004) instead of using the hypotonic lysis
7. Usually three strains within each patient are selected, even if a population of *E. coli* strains is usually expected (Darfeuille-Michaud et al., 2004; Schippa et al., 2012)
8. When dealing with viscous solutions (glycerol, DMSO, Tween-20, Triton X-100, etc.) use always low-retention 1000-μL tips (better if cut at the top, in order to enlarge the gauge) and being very slow every time when pipetting, to avoid the liquid to be retained on the tip walls. Please follow the indications provided here https://www.integra-biosciences.com/china/en/stories/how-pipette-viscous-and-volatile-liquids.
9. When preparing glycerol stocks, always use an ice bucket in order to prevent bacterial cell death due to glycerol (which acts as an ice anti-nucleant, similarly to the DMSO for eukaryotic cells). Once prepared the 5 vials, be sure to freeze them quickly. Remember to avoid repetitive and complete freeze/thawing of the vials (maximum achievable are three, in the lab experience). When a definite strain is needed, keep on ice bucket the glycerol stock and use a scalpel or a 10 μL loop under the hood in order to keep only a little piece of ice, putting it in 5 mL of liquid broth (LB or BHI or TSB, within a 15 mL tube) and let the strain grow overnight at 37 °C in agitation (200 rpm) with the cap loosely attached.
10. Change the Caco-2 cell growth medium regularly (approximately every 3 days). Composition of the growth medium: Dulbecco's

Modified Eagle's Medium—low glucose (DMEM), supplemented with 2 mmol/L glutamine, 1% Non Essential Amino Acids (NEAA), 100 IU/mL penicillin plus 0.1 mg/mL streptomycin (pen-strep solution), and 10% heat-inactivated FBS. In order to properly count Caco-2 cells please use a hemocytometer or an automated cell counter. Caco-2 cells are grown till postconfluence in order to differentiate and demonstrate the desired characteristics of small intestine enterocytes. As an indication, the doubling time of Caco-2 cells is approximately 60 h (it varies with vendor), thus in 15 days (360 h) the initial number of cells would grow by $2^6 = 64$ times. In our example, seeding $2*10^4$ cells would convey a final number of Caco-2 cells, after 15 days, equal to $1.28*10^4$. In any case, in order to properly count the mean number of Caco-2 cells after 15 days of culture, please use a hemocytometer or an automated cell counter for each experiment.

11. Multiplicity of infection (MOI) could be changed accordingly (and inversely) to the exposure time, because, from the laboratory experience and literature data, the maximum contact time among pathogenic *E. coli* (for example, AIEC) and Caco-2 cells is 6 h without noticing cell death for a MOI of 100:1 (Boudeau et al., 1999; Iebba et al., 2012; Schippa et al., 2012). Here the MOI is given as $1*10^8$ bacterial cells/$1.28*10^6$ Caco-2 cells (MOI of around 8:1, which was initially intended as MOI 10:1).

12. Washing cells is useful to detach the loosely attached bacterial cells, especially those which are not strictly adherent to the neoformed brush border and microvilli (which constitutes the postconfluence Caco-2 monolayer) (Lea, 2015)

13. The 1% concentration of Triton X-100, in deionized water, exerts no effect on bacterial viability for at least 30 min (Darfeuille-Michaud et al., 2004; Schippa et al., 2012). Eukaryotic cells are lysed by the combined action of Triton X-100 detergent nature (chemical lysis) and of deionized water (osmotic lysis).

14. For adhesion percentage, a cutoff value $\geq 0.8\%$ defines a *E.coli* strain adhesive (Darfeuille-Michaud et al., 2004; Schippa et al., 2009; Schippa et al., 2012). Report adhesion values, for a defined strain, as mean ± SEM (standard error of the mean). For reference, the mean adhesion rate on Caco-2 cells of the non-adhesive *E. coli* strain MG1655 is $0.4\% \pm 0.3\%$ of the original inoculum, whereas the adhesive strain *E. coli* EPEC 32O55 has a mean adhesion index of $2.7\% \pm 0.7\%$.

15. In order to prepare the right amount of gentamicin to be added, prepare a working solution (WS) from the mother (M). Take 100 μL of M (which is concentrated 50 mg/mL, thus 50000 μg/mL) and mix them with 900 μL of PBS, obtaining a WS of 5000 μg/mL. For each well you should use 20 μL of WS mixed with 980 μL of pre-warmed cell line medium (dilution factor 50, without penicillin and streptomycin), in order to obtain a final concentration of 100 μg/mL of gentamicin. This dilution passage is necessary to mitigate the lysing effect of the deionized water present in M, and to avoid pipetting errors (it's better to mix 20 μL than 2 μL in a final volume of 1000 μL). See the following guidelines, which summarize the best practices on pipetting: (1) https://www.hamiltoncompany.com/automated-liquid-handling/everything-you-need-to-know-about-liquid-handling/common-examples-and-best-practices; (2) https://assets.thermofisher.com/TFS-Assets/LPD/manuals/Good-Lab-Pipetting-Guide-1517630-01-Brochure-EN.pdf. For the sake of completeness, if a whole 24-well has to be taken into account, it is better to prepare a "master mix" mixing 480 μL of WS with 23.52 mL of pre-warmed cell line medium (without penicillin and streptomycin) in a 50 mL tube, then dispensing 1 mL for each well.
16. Actually the right concentration of gentamicin killing the strain used (thus the Minimum Bactericidal Concentration, MBC) should be calculated with the progressive dilution system. The Minimum Bactericidal Concentration (MBC), which is the concentration required to reduce the bacterial count by 99.99%, was determined for all strains in this study using a tissue culture medium. The antibiotic was then administered at concentrations 10 to 100 times greater than the MBC, with the MBC being less than or equal to 1 μg/mL. The concentration used here (100 μg/mL) is valid for our *E.coli* strains employed in our previous studies (especially AIEC pathovar) and in literature (Darfeuille-Michaud et al., 2004; Schippa et al., 2012).
17. Invasion percentage of a defined strain could vary depending on the eukaryotic cell line used, the incubation time (reaching or not the confluency), the bacterial exposure time (infection), and the multiplicity of infection (MOI). For example, regarding the AIEC reference strain LF82, its invasion percentages are 1.29 ± 0.55 (for Int-407cells) and 1.62 ± 0.72 (for HEp-2cells) after 3 h bacterial exposure (MOI 10:1) on seeded $2*10^4$ cells and incubation time of 15 days

(Darfeuille-Michaud et al., 2004; Schippa et al., 2012), while a previous study reported an average invasion index of 4.01% (range from 0.61 to 17.25%) after 3 h bacterial exposure (MOI 10:1) on seeded $2*10^5$ cells and incubation time of 48 h (Boudeau et al., 1999). For comparison, the non-invasive *E. coli* strain MG1655 has an invasion index of 0.001% ± 0.001% (at MOI 10:1, exposure time of 3 h, post-confluent Caco-2 cells seeded at $2*10^4$ cells/well).

18. It could be possible to observe a very high range of internalized bacteria. For example, transmission electron microscopy (TEM) highlighted till to 45 intracellular AIEC within a single epithelial cell, resembling the intracellular lifestyle of other true invasive pathogens such as *Shigella flexneri* and *Salmonella typhimurium* (Boudeau et al., 1999).

5. Concluding remarks

Assessment of adhering and invading properties is a critical aspect of understanding the pathogenic potential of *E. coli* strains and other Enterobacteriaceae. Among the different challenges and considerations in assessing adhesion and invasion there are: (i) data heterogeneity, because of variations in experimental methodologies and data presentation, which will require careful standardization; (ii) data quality, because not all studies may report all the desired parameters, leading to missing data. Standardization of data and proper reporting of experimental conditions are crucial points. These points includes: (i) converting different units of adhesion and invasion indices to a common metric (e.g., CFU/well or CFU/cell for adhesion, % invasion); (ii) ensuring consistent reporting of experimental conditions (e.g., cell type, cell contamination or "fake" cell lines—see HEp2 and Int-407 issues (Korch & Capes-Davis, 2021; Souren et al., 2022), seeding density, cell differentiation, centrifugation to enhance adhesion, MOI—multiplicity of infection, infection time, etc.). For the purposes of the present article, the Table 1 reports, as an exemplification of the huge heterogeneity in reporting adhesion/invasion results, a comprehensive literature data about the AIEC LF82 prototype from its first adhesion assessment (Darfeuille-Michaud et al., 1998) through the years. This article has highlighted the importance of assessing adhesion and invasion indexes in the context of *E. coli* pathotypes like EIEC and AIEC, comparing their strategies with those of true invasive pathogens such

Table 1 Adhesion, invasion and experimental conditions for prototypic AIEC LF82 strain.

Adhesion[a]	Invasion[b]	Cell type	Seeding/well (24-well plate)	Cell incubation (hours -h, days -d)	MOI	Centrifugation type, time, speed[c]	Infection duration	ATB type, concentration, time	References
Adhesive (Caco-2) Adhesive (Int-407)	Not performed	Caco-2 Int-407	$1*10^4$	15d (Caco-2) 48 h (Int-407)	10:1	No	3 h	Not performed	Darfeuille-Michaud et al. (1998)
Not performed	4.01% ± 3.59% ii (MOI 10:1) 1.06% ± 0.89% ii (MOI 100:1)	HEp-2	$4*10^5$	20 h	10:1 1001	No	3 h	Gentamicin, 100 µg/mL, 1 h	Boudeau, et al. (1999)
$5*10^6$ b/w inf	$1.7*10^5$ b/w inf	Int-407	$4*10^5$	20 h	10:1	No	3 h	Gentamicin, 100 µg/mL, 1 h	Boudeau, Glasser, Julien, Colombel, and Darfeuille-Michaud (2003)
4 ± 1 b/c (Caco-2) 21 ± 9 b/c (Int-407)	1.62% ± 0.72% ii (HEp-2) 1.29% ± 0.55% ii (Int-407)	Caco-2 Int-407 HEp-2	$2*10^4$ (Caco-2) $4*10^5$ (Int-407, HEp-2)	15d (Caco-2) 20 h (Int-407, HEp-2)	10:1	No	3 h	Gentamicin, 100 µg/mL, 1 h	Darfeuille-Michaud et al. (2004)
2.06 ± 0.94 $*10^4$ b/w (Int-407) 0.45 ± 0.33 $*10^4$ b/w (HT29)	12.90% ± 38.99% ii 12.36% ± 29.87% ii	Int-407 HT29	$4*10^5$	To confluence	10:1	No	3 h	Gentamicin, 100 µg/mL, 1 h	Martin et al. (2004)
186.6 % ± 63.9 % ii	1.10% ± 0.28% ii (w/o ctrfg) 7.99% ± 2.98% ii (w ctrfg)	HEp-2	$4*10^5$	20 h	10:1	Inv, 10 m, 2000 *g*	3 h	Gentamicin, 100 µg/mL, 1 h	Boudeau, Barnich, and Darfeuille-Michaud (2001)

$2.9 \pm 1.4 * 10^6$ b/w $1.6 \pm 0.2 * 10^7$ b/w $4.8 \pm 0.1 * 10^6$ b/w $3.7 \pm 0.5 * 10^6$ b/w	$1.5 \pm 0.9 * 10^4$ b/w $4.2 \pm 0.9 * 10^5$ b/w $3.9 \pm 1.6 * 10^5$ b/w $2.3 \pm 0.6 * 10^4$ b/w	Int-407	$4*10^5$	20 h	10:1	No	3 h	Gentamicin, 100 μg/mL, 1 h	Rolhion, Carvalho, and Darfeuille-Michaud (2007)
$240*10^6$ b/w (Caco-2) inf $160*10^6$ b/w (T84) inf	$510*10^2$ b/w (Caco-2) inf $390*10^2$ b/w (T84) inf	Caco-2 T84	Unknown	5-6d	100:1	No	3 h	Gentamicin, 50 μg/mL, 1 h	Eaves-Pyles et al. (2008)
Not performed	1.29% ii	HEp-2	Unknown	To confluence	10:1	No	3 h	Gentamicin, 100 μg/mL, 1 h	Negroni et al. (2011)
4.25 b/c inf	Not performed	T84	$2*10^5$	48 h	10:1	Adh, 10 m, 900 *g*	30 m	Not performed	Negroni et al. (2011); Dreux et al. (2013)
Not reported	0.95% ±0.3% ii	HEp-2	$1*10^5$ (adh) $2*10^5$ (inv)	24 h	10:1	No	3 h	Gentamicin, 100 μg/mL, 1 h	Conte et al. (2014)
10 % ii	1.5% ii	HEp-2	$1*10^5$	48 h	10:1	Adh/Inv, 2.5 m, 500 *g*	2 h	Gentamicin, 100 μg/mL, 1 h	Conte et al. (2016)
62.83 ± 5.08 b/c	$12.23\% \pm 2.01\%$ ii	Int-407	$4*10^5$	20 h	10:1	No	3 h	Gentamicin, 100 μg/mL, 1 h	O'Brien et al. (2017)
25.66 ± 15.7 b/c	$2.26\% \pm 1.349\%$ ii	Int-407	$4*10^5$	20 h	10:1	No	3 h	Gentamicin, 100 μg/mL, 1 h	Camprubí-Font et al. (2018)
$5.13*10^6$ b/w	1.12% ii	Caco-2	$2*10^5$	Unknown	10:1	No	3 h	Gentamicin, 100 μg/mL, 1 h	Kittana et al. (2023)

[a]b/w: bacteria per well, b/c: bacterial per cell; ii: initial inoculum; inf: inferred from graphs.
[b]b/w: bacteria per well; ii: initial inoculum; inf: inferred from graphs; w/o ctrfg: without centrifugation; w ctrfg: with centrifugation.
[c]Centrifugation of infected cells. Adh/Inv: for adhesion and invasion; Adh: for adhesion; Inv: for invasion.

as *Shigella* and *Salmonella*. By deciphering the specific adhesion and invasion mechanisms employed by these pathogens, starting from the assessment of their adhesive/invasive indexes, we can pave the way for improved diagnostic methods, targeted therapeutic interventions, and ultimately, better management of enteric infections. It is worth noting how specific FimH blockers are currently used to hinder or block adhesion of *E.coli* AIEC strains to ameliorate Crohn's disease patients intestinal inflammation (Chevalier et al., 2021).

References

Arizmendi, O., Picking, W. D., & Picking, W. L. (2016). Macrophage apoptosis triggered by IpaD from *Shigella flexneri*. *Infection and Immunity*, *84*, 1857–1865.

Barbau-Piednoir, E., Denayer, S., Botteldoorn, N., Dierick, K., De Keersmaecker, S. C. J., & Roosens, N. H. (2018). Detection and discrimination of five *E. coli* pathotypes using a combinatory SYBR® Green qPCR screening system. *Applied Microbiology and Biotechnology*, *102*, 3267–3285.

Bonet-Rossinyol, Q., Camprubí-Font, C., López-Siles, M., & Martinez-Medina, M. (2023). Identification of differences in gene expression implicated in the Adherent-Invasive phenotype during infection of intestinal epithelial cells. *Frontiers in Cellular and Infection Microbiology*, *13*, 1228159.

Boudeau, J., Barnich, N., & Darfeuille-Michaud, A. (2001). Type 1 pili-mediated adherence of *Escherichia coli* strain LF82 isolated from Crohn's disease is involved in bacterial invasion of intestinal epithelial cells. *Molecular Microbiology*, *39*, 1272–1284.

Boudeau, J., Glasser, A.-L., Julien, S., Colombel, J.-F., & Darfeuille-Michaud, A. (2003). Inhibitory effect of probiotic *Escherichia coli* strain Nissle 1917 on adhesion to and invasion of intestinal epithelial cells by adherent–invasive *E. coli* strains isolated from patients with Crohn's disease. *Alimentary Pharmacology & Therapeutics*, *18*, 45–56.

Boudeau, J., Glasser, A. L., Masseret, E., Joly, B., & Darfeuille-Michaud, A. (1999). Invasive ability of an *Escherichia coli* strain isolated from the ileal mucosa of a patient with Crohn's disease. *Infection and Immunity*, *67*, 4499–4509.

Buc, E., Dubois, D., Sauvanet, P., Raisch, J., Delmas, J., Darfeuille-Michaud, A., et al. (2013). High prevalence of mucosa-associated *E. coli* producing cyclomodulin and genotoxin in colon cancer. *PLoS One*, *8*, e56964.

Camprubí-Font, C., Bustamante, P., Vidal, R. M., O'Brien, C. L., Barnich, N., & Martinez-Medina, M. (2020). Study of a classification algorithm for AIEC identification in geographically distinct *E. coli* strains. *Scientific Reports*, *10*, 8094.

Camprubí-Font, C., Ewers, C., Lopez-Siles, M., & Martinez-Medina, M. (2019). Genetic and phenotypic features to screen for putative adherent-invasive. *Frontiers in Microbiology*, *10*, 108.

Camprubí-Font, C., Lopez-Siles, M., Ferrer-Guixeras, M., Niubó-Carulla, L., Abellà-Ametller, C., Garcia-Gil, L. J., et al. (2018). Comparative genomics reveals new single-nucleotide polymorphisms that can assist in identification of adherent-invasive *Escherichia coli*. *Scientific Reports*, *8*, 2695.

Camprubí-Font, C., & Martinez-Medina, M. (2020). Why the discovery of adherent-invasive molecular markers is so challenging? *World Journal of Biological Chemistry*, *11*, 1–13.

Céspedes, S., Saitz, W., Del Canto, F., De la Fuente, M., Quera, R., Hermoso, M., et al. (2017). Genetic diversity and virulence determinants of strains isolated from patients with Crohn's disease in Spain and Chile. *Frontiers in Microbiology*, *8*, 639.

Chevalier, G., Laveissière, A., Desachy, G., Barnich, N., Sivignon, A., Maresca, M., et al. (2021). Blockage of bacterial FimH prevents mucosal inflammation associated with Crohn's disease. *Microbiome*, *9*, 176.

Clements, A., Berger, C. N., Lomma, M., & Frankel, G. (2013). *Escherichia coli: Chapter 15. Type 3 secretion effectors*. Elsevier Inc (Chapters).

Conte, M. P., Aleandri, M., Marazzato, M., Conte, A. L., Ambrosi, C., Nicoletti, M., et al. (2016). The adherent/invasive *Escherichia coli* strain LF82 invades and persists in human prostate cell line RWPE-1, activating a strong inflammatory response. *Infection and Immunity*. https://doi.org/10.1128/iai.00438-16.

Conte, M. P., Longhi, C., Marazzato, M., Conte, A. L., Aleandri, M., Lepanto, M. S., et al. (2014). Adherent-invasive *Escherichia coli* (AIEC) in pediatric Crohn's disease patients: phenotypic and genetic pathogenic features. *BMC Research Notes*, 7, 1–12.

Costa, R. F. A., Ferrari, M. L. A., Bringer, M.-A., Darfeuille-Michaud, A., Martins, F. S., & Barnich, N. (2020). Characterization of mucosa-associated *Escherichia coli* strains isolated from Crohn's disease patients in Brazil. *BMC Microbiology*, *20*, 178.

Croxen, M. A., Law, R. J., Scholz, R., Keeney, K. M., Wlodarska, M., & Finlay, B. B. (2013). Recent advances in understanding enteric pathogenic *Escherichia coli*. *Clinical Microbiology Reviews*, *26*, 822–880.

Darfeuille-Michaud, A., Boudeau, J., Bulois, P., Neut, C., Glasser, A.-L., Barnich, N., et al. (2004). High prevalence of adherent-invasive *Escherichia coli* associated with ileal mucosa in Crohn's disease. *Gastroenterology*, *127*, 412–421.

Darfeuille-Michaud, A., Neut, C., Barnich, N., Lederman, E., Di Martino, P., Desreumaux, P., et al. (1998). Presence of adherent Escherichia coli strains in ileal mucosa of patients with Crohn's disease. *Gastroenterology*, *115*, 1405–1413.

Deshpande, N. P., Wilkins, M. R., Mitchell, H. M., & Kaakoush, N. O. (2015). Novel genetic markers define a subgroup of pathogenic *Escherichia coli* strains belonging to the B2 phylogenetic group. *FEMS Microbiology Letters*, *362*.

Desilets, M., Deng, X., Rao, C., Ensminger, A. W., Krause, D. O., Sherman, P. M., et al. (2016). Genome-based definition of an inflammatory bowel disease-associated adherent-invasive *Escherichia coli* pathovar. *Inflammatory Bowel Diseases*, *22*, 1–12.

Di Pasquale, P., Caterino, M., Di Somma, A., Squillace, M., Rossi, E., Landini, P., et al. (2016). Exposure of *E. coli* to DNA-methylating agents impairs biofilm formation and invasion of eukaryotic cells via down regulation of the N-acetylneuraminate lyase NanA. *Frontiers in Microbiology*, 7, 147.

Dreux, N., Denizot, J., Martinez-Medina, M., Mellmann, A., Billig, M., Kisiela, D., et al. (2013). Point mutations in FimH Adhesin of Crohn's disease-associated adherent-invasive *Escherichia coli* enhance intestinal inflammatory response. *PLoS Pathogens*, *9*, e1003141.

Du, J., Reeves, A. Z., Klein, J. A., Twedt, D. J., Knodler, L. A., & Lesser, C. F. (2016). The type III secretion system apparatus determines the intracellular niche of bacterial pathogens. *Proceedings of the National Academy of Sciences of the United States of America*, *113*, 4794–4799.

Eaves-Pyles, T., Allen, C. A., Taormina, J., Swidsinski, A., Tutt, C. B., Jezek, G. E., et al. (2008). *Escherichia coli* isolated from a Crohn's disease patient adheres, invades, and induces inflammatory responses in polarized intestinal epithelial cells. *International Journal of Medical Microbiology*, *298*, 397–409.

Elhenawy, W., Oberc, A., & Coombes, B. K. (2018). A polymicrobial view of disease potential in Crohn's-associated adherent-invasive *E. coli*. *Gut Microbes*, *9*, 166–174.

Eng, S.-K., Pusparajah, P., Ab Mutalib, N.-S., Ser, H.-L., Chan, K.-G., & Lee, L.-H. (2015). Salmonella: A review on pathogenesis, epidemiology and antibiotic resistance. *Front Life Sciences*, *8*, 284–293.

Falkow, S., Small, P., Isberg, R., Hayes, S. F., & Corwin, D. (1987). A molecular strategy for the study of bacterial invasion. *Reviews of Infectious Diseases*, *9*(Suppl 5), S450–S455.

Garcia-Angulo, V. A., Farfan, M. J., & Torres, A. G. (2013). *Escherichia coli*: Chapter 11. In *Hybrid and potentially pathogenic Escherichia coli strains*. Chapters: Elsevier Inc.

Glasser, A. L., Boudeau, J., Barnich, N., Perruchot, M. H., Colombel, J. F., & Darfeuille-Michaud, A. (2001). Adherent invasive *Escherichia coli* strains from patients with Crohn's disease survive and replicate within macrophages without inducing host cell death. *Infection and Immunity*, *69*, 5529–5537.

Gomes, T. A. T., Dobrindt, U., Farfan, M. J., & Piazza, R. M. (2021). *Interaction of Pathogenic Escherichia coli with the Host: Pathogenomics*. Frontiers Media SA: Virulence and Antibiotic Resistance.

Hernandez, L. D., Pypaert, M., Flavell, R. A., & Galán, J. E. (2003). A Salmonella protein causes macrophage cell death by inducing autophagy. *The Journal of Cell Biology*, *163*, 1123–1131.

Iebba, V., Conte, M. P., Lepanto, M. S., Di Nardo, G., Santangelo, F., Aloi, M., et al. (2012). Microevolution in fimH gene of mucosa-associated *Escherichia coli* strains isolated from pediatric patients with inflammatory bowel disease. *Infection and Immunity*, *80*, 1408–1417.

Jajere, S. M. (2019). A review of with particular focus on the pathogenicity and virulence factors, host specificity and antimicrobial resistance including multidrug resistance. *Veterinary World*, *12*, 504–521.

Johnson, D. I. (2017). *Bacterial Pathogens and Their Virulence Factors*. Springer.

Kaper, J. B., Nataro, J. P., & Mobley, H. L. (2004). Pathogenic *Escherichia coli*. *Nature Reviews. Microbiology*, *2*, 123–140.

Keita, Å. V., Alkaissi, L. Y., Holm, E. B., Heil, S. D. S., Chassaing, B., Darfeuille-Michaud, A., et al. (2020). Enhanced *E. coli* LF82 translocation through the follicle-associated epithelium in Crohn's disease is dependent on long polar fimbriae and CEACAM6 expression, and increases paracellular permeability. *Journal of Crohn's & Colitis*, *14*, 216–229.

Kittana, H., Gomes-Neto, J. C., Heck, K., Juritsch, A. F., Sughroue, J., Xian, Y., et al. (2023). Evidence for a causal role for *Escherichia coli* Strains Identified as Adherent-Invasive (AIEC) in intestinal inflammation. *mSphere*, *8*, e0047822.

Korch, C. T., & Capes-Davis, A. (2021). The extensive and expensive impacts of HEp-2 [HeLa], Intestine 407 [HeLa], and other false cell lines in journal publications. *SLAS Discovery*, *26*, 1268–1279.

Kotloff, K. L., Nataro, J. P., Blackwelder, W. C., Nasrin, D., Farag, T. H., Panchalingam, S., et al. (2013). Burden and aetiology of diarrhoeal disease in infants and young children in developing countries (the Global Enteric Multicenter Study, GEMS): a prospective, case-control study. *Lancet*, *382*, 209–222.

Lampel, K. A., Formal, S. B., & Maurelli, A. T. (2018). A Brief History of. *EcoSal Plus*, *8*.

Law, R. J., Gur-Arie, L., Rosenshine, I., & Finlay, B. B. (2013). In vitro and in vivo model systems for studying enteropathogenic *Escherichia coli* infections. *Cold Spring Harbor Perspectives in Medicine*, *3*, a009977.

Lea, T. (2015). Caco-2 cell line. In K. Verhoeckx, P. Cotter, I. López-Expósito, C. Kleiveland, T. Lea, A. Mackie, ... H. Wichers (Eds.), *The Impact of Food Bioactives on Health: in vitro and ex vivo models*. Cham, CH: Springer.

Lee, J. G., Han, D. S., Jo, S. V., Lee, A. R., Park, C. H., Eun, C. S., et al. (2019). Characteristics and pathogenic role of adherent-invasive *Escherichia coli* in inflammatory bowel disease: Potential impact on clinical outcomes. *PLoS One*, *14*, e0216165.

Marchi, S., Morroni, G., Pinton, P., & Galluzzi, L. (2022). Control of host mitochondria by bacterial pathogens. *Trends in Microbiology*, *30*, 452–465.

Martin, H. M., Campbell, B. J., Hart, C. A., Mpofu, C., Nayar, M., Singh, R., et al. (2004). Enhanced *Escherichia coli* adherence and invasion in Crohn's disease and colon cancer. *Gastroenterology*, *127*, 80–93.

Martinez-Medina, M., Aldeguer, X., Lopez-Siles, M., González-Huix, F., López-Oliu, C., Dahbi, G., et al. (2009). Molecular diversity of *Escherichia coli* in the human gut: new ecological evidence supporting the role of adherent-invasive *E. coli* (AIEC) in Crohn's disease. *Inflammatory Bowel Diseases*, *15*, 872–882.

Martinez-Medina, M., & Garcia-Gil, L. J. (2014). *Escherichia coli* in chronic inflammatory bowel diseases: An update on adherent invasive *Escherichia coli* pathogenicity. *World Journal of Gastrointestinal Pathophysiology*, *5*, 213–227.

Mattock, E., & Blocker, A. J. (2017). How do the virulence factors of work together to cause disease? *Frontiers in Cellular and Infection Microbiology*, *7*, 64.

Mayorgas, A., Dotti, I., Martínez-Picola, M., Esteller, M., Bonet-Rossinyol, Q., Ricart, E., et al. (2021). A novel strategy to study the invasive capability of adherent-invasive by using human primary organoid-derived epithelial monolayers. *Frontiers in Immunology*, *12*, 646906.

Nash, J. H., Villegas, A., Kropinski, A. M., Aguilar-Valenzuela, R., Konczy, P., Mascarenhas, M., et al. (2010). Genome sequence of adherent-invasive *Escherichia coli* and comparative genomic analysis with other *E. coli* pathotypes. *BMC Genomics*, *11*, 667.

Negroni, A., Costanzo, M., Vitali, R., Superti, F., Bertuccini, L., Tinari, A., et al. (2011). Characterization of adherent-invasive *Escherichia coli* isolated from pediatric patients with inflammatory bowel disease. *Inflammatory Bowel Diseases*, *18*, 913–924.

Nguyen, H. T. T., Dalmasso, G., Müller, S., Carrière, J., Seibold, F., & Darfeuille-Michaud, A. (2014). Crohn's disease-associated adherent invasive Escherichia coli modulate levels of microRNAs in intestinal epithelial cells to reduce autophagy. *Gastroenterology*, *146*, 508–519.

Nougayrède, J.-P., & Donnenberg, M. S. (2004). Enteropathogenic *Escherichia coli* EspF is targeted to mitochondria and is required to initiate the mitochondrial death pathway. *Cellular Microbiology*, *6*, 1097–1111.

O'Brien, C. L., Bringer, M.-A., Holt, K. E., Gordon, D. M., Dubois, A. L., Barnich, N., et al. (2017). Comparative genomics of Crohn's disease-associated adherent-invasive. *Gut*, *66*, 1382–1389.

Pakbin, B., Brück, W. M., & Rossen, J. W. A. (2021). Virulence factors of enteric pathogenic: A review. *International Journal of Molecular Sciences*, *22*.

Pinaud, L., Sansonetti, P. J., & Phalipon, A. (2018). Host cell targeting by enteropathogenic bacteria T3SS effectors. *Trends in Microbiology*, *26*, 266–283.

Quainoo, S., Coolen, J. P. M., van Hijum, S. A. F. T., Huynen, M. A., Melchers, W. J. G., van Schaik, W., et al. (2017). Whole-genome sequencing of bacterial pathogens: the future of nosocomial outbreak analysis. *Clinical Microbiology Reviews*, *30*, 1015–1063.

Ramachandran, R. P., Spiegel, C., Keren, Y., Danieli, T., Melamed-Book, N., Pal, R. R., et al. (2020). Mitochondrial targeting of the enteropathogenic *Escherichia coli* map triggers calcium mobilization, ADAM10-MAP kinase signaling, and host cell apoptosis. *MBio*, *11*.

Rivas, L., Mellor, G. E., Gobius, K., & Fegan, N. (2015). *Detection and Typing Strategies for Pathogenic Escherichia coli*. Springer.

Robins-Browne, R. M., Holt, K. E., Ingle, D. J., Hocking, D. M., Yang, J., & Tauschek, M. (2016). Are Pathotypes Still Relevant in the Era of Whole-Genome Sequencing? *Frontiers in Cellular and Infection Microbiology*, *6*, 141.

Rolhion, N., Barnich, N., Bringer, M.-A., Glasser, A.-L., Ranc, J., Hébuterne, X., et al. (2010). Abnormally expressed ER stress response chaperone Gp96 in CD favours adherent-invasive *Escherichia coli* invasion. *Gut*, *59*, 1355–1362.

Rolhion, N., Carvalho, F. A., & Darfeuille-Michaud, A. (2007). OmpC and the σE regulatory pathway are involved in adhesion and invasion of the Crohn's disease-associated *Escherichia coli* strain LF82. *Molecular Microbiology, 63*, 1684–1700.

Rolhion, N., Hofman, P., & Darfeuille-Michaud, A. (2011). The endoplasmic reticulum stress response chaperone: Gp96, a host receptor for Crohn disease-associated adherent-invasive *Escherichia coli*. *Gut Microbes, 2*, 115–119.

Schippa, S., Conte, M. P., Borrelli, O., Iebba, V., Aleandri, M., Seganti, L., et al. (2009). Dominant genotypes in mucosa-associated *Escherichia coli* strains from pediatric patients with inflammatory bowel disease. *Inflammatory Bowel Diseases, 15*, 661–672.

Schippa, S., Iebba, V., Totino, V., Santangelo, F., Lepanto, M., Alessandri, C., et al. (2012). A potential role of *Escherichia coli* pathobionts in the pathogenesis of pediatric inflammatory bowel disease. *Canadian Journal of Microbiology, 58*, 426–432.

Sharma, A., & Puhar, A. (2019). Gentamicin protection assay to determine the number of intracellular bacteria during infection of human TC7 intestinal epithelial cells by. *Bio-Protocol, 9*, e3292.

Souren, N. Y., Fusenig, N. E., Heck, S., Dirks, W. G., Capes-Davis, A., Bianchini, F., et al. (2022). Cell line authentication: a necessity for reproducible biomedical research. *The EMBO Journal, 41*, e111307.

Srikanth, C. V., Mercado-Lubo, R., Hallstrom, K., & McCormick, B. A. (2011). Salmonella effector proteins and host-cell responses. *Cellular and Molecular Life Sciences, 68*, 3687–3697.

Swidsinski, A., Loening-Baucke, V., Theissig, F., Engelhardt, H., Bengmark, S., Koch, S., et al. (2007). Comparative study of the intestinal mucus barrier in normal and inflamed colon. *Gut, 56*, 343–350.

The, H. C., Thanh, D. P., Holt, K. E., Thomson, N. R., & Baker, S. (2016). The genomic signatures of Shigella evolution, adaptation and geographical spread. *Nature Reviews. Microbiology, 14*, 235–250.

Tiku, V., Tan, M.-W., & Dikic, I. (2020). Mitochondrial functions in infection and immunity. *Trends in Cell Biology, 30*, 263–275.

van den Beld, M. J. C., & Reubsaet, F. A. G. (2012). Differentiation between Shigella, enteroinvasive *Escherichia coli* (EIEC) and noninvasive *Escherichia coli*. *European Journal of Clinical Microbiology & Infectious Diseases, 31*, 899–904.

Vashisht, V., Vashisht, A., Mondal, A. K., Farmaha, J., Alptekin, A., Singh, H., et al. (2023). Genomics for emerging pathogen identification and monitoring: Prospects and obstacles. *BioMedInformatics, 3*, 1145–1177.

Verma, P., Gandhi, S., Lata, K., & Chattopadhyay, K. (2021). Pore-forming toxins in infection and immunity. *Biochemical Society Transactions, 49*, 455–465.

Worrall, L. J., Bergeron, J. R. C., & Strynadka, N. C. J. (2013). *Escherichia coli: Chapter 14. Type 3 secretion systems*. Elsevier Inc (Chapters).

Zangara, M. T., Darwish, L., & Coombes, B. K. (2023). Characterizing the pathogenic potential of Crohn's disease-associated adherent-invasive. *EcoSal Plus, 11*. eesp00182022.

Zhang, Y., Rowehl, L., Krumsiek, J. M., Orner, E. P., Shaikh, N., Tarr, P. I., et al. (2015). Identification of candidate adherent-invasive *E. coli* signature transcripts by genomic/transcriptomic analysis. *PLoS One, 10*, e0130902.

Zhou, G., Zhao, Y., Ma, Q., Li, Q., Wang, S., & Shi, H. (2023). Manipulation of host immune defenses by effector proteins delivered from multiple secretion systems of and its application in vaccine research. *Frontiers in Immunology, 14*, 1152017.

Printed in the United States
by Baker & Taylor Publisher Services